D0594920

BE *Pretty* GET *Married*

AND ALWAYS DRINK TAB

A MEMOIR

GIGI ANDERS

rayo

An Imprint of HarperCollinsPublishers

HarperCollins books may be purchased for educational, business, or sales promotional use. For information, please write: Special Markets Department, HarperCollins Publishers Inc., 10 East 53rd Street, New York, NY 10022.

Originally published in hardcover by HarperCollins in 2005 under the title JUBANA!

FIRST EDITION

Designed by Gretchen Achilles

Library of Congress Cataloging-in-Publication Data

Anders, Gigi.
    Jubana! : the awkwardly true and dazzling adventures of a Jewish Cubana goddess / Gigi Anders—1st ed.
        p.   cm.
    ISBN-10: 0-06-056370-2
    ISBN-13: 978-0-06-056370-7
    1. Anders, Gigi. 2. Cuban American women—Biography. 3. Jewish women—United States—Biography. 4. Cuban Americans—Biography. 5. Jews, Cuban—United States—Biography. 6. Anders, Gigi—Family. 7. Jews, Cuban—Biography. 8. Jews—Cuba—Biography. I. Title.

E184.C97A53 2005
973'.04924'0092—dc22   2004062044
[B]

06   07   08   09   DIX/RRD   10   9   8   7   6   5   4   3   2   1

To the memories of my Jubana baby sister, Cecilia,
and to Valerie, my North American WASP cousin who saved me.
Con cariño.

BE
Pretty
GET
Married
AND ALWAYS DRINK TAB

CHAPTER ONE

Guerrilla Baby

*It's the classic Latina position: Be pretty, get married, and shut the fuck up.

I am not a classic Latina.

I am a Jubana, a Cuban Jewess.

And when you're a *bride-to-be* Jubana, you have to know you're heading straight into the mondo bizarro jaws of cross-cultural hell. Especially if, like me, you're an only child (which I am, except for my two American-born younger brothers). My mother, Ana, also a Cuban-born only daughter with two brothers, was treated by her indulgent parents like the quintessential Jewish Cuban Princess (JCP) she was and would always be, Fidel Castro's revolution be damned. The princess royal's parents, Boris and Dora, had emigrated as destitute teenagers from Russia and Lithuania to Cuba in the early 1920s. And just like my Polish-born paternal grandparents, Leon and Zelda, they spoke Yiddish and Hebrew with Cuban accents, Spanish with Yiddish accents, and English with Yiddish-Cuban accents.

Boris, born Boruch Benes, was a self-made man and Reform

Jew. He started out selling handkerchiefs, bolts of lace, and fabrics, and eventually became the wildly prosperous owner of Camisetas Perro (literally translated, Dog Undershirts—it sounds way better in *Español*), sort of the Victoria's (and Victor's) Secret of its day. He and Dora threw their only daughter the grandest marital bash of that winter season. At my mother's 1954 December wedding in Havana there were 750 guests. That's *muchos* silk undies. (Think wedding in *Goodbye, Columbus*, only everyone sounded like Ricky Ricardo or Ricky Ricardo with a Yiddish accent.) Mami's only job on that day was to show up in perfect makeup; a heavy, white, hand-embroidered velvet dress; smile; and do whatever she was told. Which she did. She agreed to have virtually zero input but her attractive presence and choice in groom.

But I've examined Mami's hand-tinted bridal portrait in my parents' Silver Spring, Maryland, living room. I know what really lies behind the twenty-one-year-old bride's crimson-colored smile.

"I'm goheengh to get joo, sohkehr."

The "joo" would be . . . who? My father? Well, that was a given. My father, David, has never been able to say the word *no* to my mother. Indeed, that was a very strong selling point to get him on her short list. Because in case of doubt, worship works on JCPs.

Was *I* the sohkehr she was gonna get? Probably, though I wasn't born yet. Mami always said that until she had me, she could wear bikinis. Thanks to me, she, who was otherwise beautiful and perfect, was *deesfeegur-ed* with ugly, permanent stretch marks, and forever relegated to maillots.

That is hard-core guilt. That is the classic Jewish way. Be alive and be guilty—over what exactly, no one knows and it really doesn't matter anyway. Just be it.

As a result of the disfigurement and due to the presence of my vulva instead of (the infinitely more desirable, powerful, valuable, and superior) penis, I was subjected—just as Mami had been back in her day—to control and guilt, the respective Latin and Jewish mega-bullies. Not that any of us are bitter or anything.

Now, intellectually, we all in my family realize we've been out of Cuba, our homeland, for well over forty years. We understand things have, you know, changed. Today's typical bride is well past twenty-one and is the primary choreographer of her own damn wedding. The parents' primary contribution is to pay for some or all of it and to consider that payment a gift.

Under normal circumstances, with at least seminormal parents (i.e., parents who aren't the children of Cuban Jews and didn't experience in their own lives yet another generation of political/emotional/geographic dislocation at a tender age, and who, as a result, are terminally nuts), "gift" would mean, uh, "gift." But in my case, the planning and execution of my wedding is an all-out conflict, an estrogen-espresso-propulsed struggle for power and control.

I'm sitting with my parents in their family room, going over the guest list. My fiancé has wisely chosen to be suddenly indisposed elsewhere in the country. Neither his relatives nor any of our respective friends' names appear on that list. Mami's going for a Kool (she's the only white person I know of who smokes that brand) and her five millionth cup of jet fuel, aka espresso. Papi is spaced out on the couch, absorbed in the Redskins' latest near-perfect losing streak.

"Dahveed!" Mami commands Papi. "Come over here an' look at our lees. We wan' joor eenpoot."

"No we don't," I tell her, lighting a Parliament and reaching for a TaB. So delicious and soothing, this ritual. My vegetable and carbonated-water diet has kept me going for a good thirty years. I consider it a religion, really. I've actually turned down jobs and spurned relationships because they weren't located near TaB access.

Mami scowls at me.

"Dad," I continue, exhaling, "couldn't care less."

"Dad kehrz! He kehrz a LOT. Dahveed! Show joor daughter joo KEHR."

Dad looks up wanly and reluctantly joins us.

"Now look," Mami says, pushing the huge list in front of him. It's on a legal pad she stole from work. Mami isn't one to "buy" things. Actually, she resents having to pay for anything. She believes she should be exempt because Castro took everything away from her. Therefore, she's special. Very special. Castro made her an exiled victim, and she's pissed about it. Cubanly pissed. The kind of pissed you don't get over. And that's why she feels entitled to steal. People who pay for things, like people who voluntarily slow down at yellow lights, are "total sohkehrz."

Papi barely glances at the list. I know what he's thinking:

I love Gigi but my Redskins are on and this is girlie shit and I'm gonna wind up having to sell one or both of my huevos to finance this fucking fiesta. But if I don't, I'll feel too guilty to live. Why do broads make such a fuss over—Hey! Was that OUR touchdown or theirs?!?

"Dahveed! Where are joo goheengh?"

Dad's back on the couch, body coiled, fist in mouth, remote at the ready.

"Mom, he wants to watch the game. Forget it, okay? Now let's get back to this list. What about my friends? What about my man's friends and family? I don't see them on here. That's at least another fifty, sixty guests right there."

"What friends? Hees who? Anyway, dat ees sooo totally boreengh. Less move ON."

She pulls out a fresh sheet of paper.

"Okay, less focus on de foo'. I lohv stahrteengh weeth gehfeel-teh feeshy."

"Are you insane? We're Reform. We are not having gefilte fish as an appetizer. I'm sorry, it's gross. Gefilte fish is . . . *gray*. It's suspended in *gelatin*. It comes in a *jar*. It's like Fish Food, like, the Mystery Fish. You know how, like, at the supermarket they have that cheese called 'Cheese Food'? That's what this is."

As usual, my mother has absolutely no idea what I'm talking about and even less interest. The woman does not believe in groceries. I don't mean she steals food at the supermarket. I mean she does not believe in groceries. As long as she's got her espresso, Kools, and a table in the smoking section at the Cheesecake Factory, she's good to go.

"I lohv gehfeelteh feeshy," Mami continues, oblivious to all but her own preferences. "Ees dehleeshohs. Ees one of my favoreet foo's. Actually, I weesh I ate eet all jear aroun', ees so goo'."

For the nine billionth time, I realize that I, like poor dead JFK during the misbegotten 1961 Bay of Pigs invasion, am deluded to think I have any real chance of reaching compromise, much less victory over the dictator. Yet I, designated sohkehr, foolishly press on, thus ensuring I'll be got. Just like my fellow Cuban refugee sohkehrz who were haplessly overwhelmed and killed by Castro's army. The lucky ones got taken prisoner and had to drink their own urine to stay alive.

"I was thinking," I said, reaching for another TaB, "that since this is a Jewish AND Cuban wedding, it's key to serve typically Cuban foods. Tell you what. I'll give you—uch—gefilte fish if you'll let me have shrimp in *mojito* [garlic, onion, and olive oil] sauce and *croquetas de jamón* [ham croquettes]."

"Are joo KREHSEE?" she cried. "Chehlfeesh and pork at a Je-weesh weddeengh? Weeth de RABBI der? *Estás loca?*" Are you crazy?

"Why? We're REFORM Jew Cubans. I mean, hello. This is *whose* wedding?"

"Mine! Ees all mine, mine, mine! *El que paga, manda.* Whoever has de money has de power!"

Since I am a Jubana, my culturally imposed and sole raison d'être is to replicate all things Mami, who is deemed the archetype of Jewish and Latina femininity. However, I personally cannot recall a single instance in my youth when I fantasized about my wedding day. Not the dress, not the ring, not the groom, *nada.* Obviously, that meant there was something wrong with me. I was interested in books, poetry, writing, animals, TaB, fashion and celebrity magazines, TV, music, cooking, cigarettes, and movies. But Mami *had* imagined that magical conjugal scenario for me, right down to the gehfeelteh feeshy, over and over from the moment of my conception—one of the only times the woman has ever planned ahead. That was what her mother had done with her, and her mother before her, and so on and so on, all the way to Creation.

My point, and I swear to God I have one, is that there are really only three things you can do with a Jubana baby daughter, and these were all premeditated while I was still floating blissfully unaware in Mami's womb:

1. Control Gigi's life and appearance (somebody's got to).

2. Plan out Gigi's wedding (a party for Mami and her 750 closest friends!).

3. Figure out how to afterward cram said 750 into *la sala,* the living room, for the baby shower(s).

That's the drill, that's the deal, that's the traditional Juban Way. The only way for a girl.

That's why I'm here.

Verdad?

You might think that by now assimilation would have smoothed out all my Jubanity. Honey, I've tried melting into that so-called melting pot. Nothing happened. I stuck out like an un-killable pig bone in a stew of mushy black beans, or frizz-prone hair on a rainy day. As Mami Dearest says, "Honey, joo could put a paper bag over joor hayt an' guess what? Jood steel be con-speekuohs."

Life has its rules, and the first one I learned is that it's never too early to worry about your appearance. How else are you gonna get some high-income sohkehr to marry you? So this is what you need to understand about Juban girl babyhood: You learn pretty much everything you need to know about life long before your third birthday. For starters, you automatically have your ears pierced before you ever leave the maternity ward with your mother. They give you about one carefree postbirth day of infant fun and then—PRICK. It's the female equivalent of a bris. There is no such thing as a Jubana baby without pierced ears bearing tiny pearls/diamonds/gold balls. That teaches us gals that beauty equals pain and you'll have many different kinds and degrees of it for the rest of your life.

For even the humblest of Jubanas—and trust me, humility is not a salient feature of our personality profile—being obsessively meticulous about beauty is a given. Beauty is never frivolous. In our scheme of priorities, lipstick always trumps, like, food. About that, Mami Dearest would say, "Das so ohbveeohs, ees rehdeecoolohs." What some unfortunate non-Jubanas might consider self-indulgent—understanding the concept of and need for bidets; eschewing hair coloring you can buy at the drugstore; and always wearing good perfume on your body AND in your hair (even if you're alone)—is for us imperative.

That last item can start right at birth, sometimes even BEFORE the ear piercing, depending on a Jubana baby's hair presence or lack thereof. When I was born, I emerged into La Habana's pretty world and what would have been a charmed life, had Feedehl Castro not happened, with an abnormally huge head (it runs in the family) full of thick auburn hair with golden-red highlights. This was a source of intense pride and excitement for my glamorous redheaded Mami, who considers any hint of red in hair ohbveeohsly superior. She sent for the Agua de Violetas at once.

Hair is good. Hair with Agua de Violetas in it is better. Agua de Violetas. Violet water. A topaz-colored eau de cologne for the hair. As Dora had done for her, Mami anointed me with the lovely solution that streamed in cool rivulets down my scalp, flushing the air with the refreshing, comforting fragrance of cut oranges and fresh violets.

When Mami and her girlfriends were growing up in the forties and fifties, they all wore it, and when they had their babies, they'd give bottles of Violetas as shower gifts. Piled on the mothers' beds were the rectangular boxes containing the elegant bottles, the "Fragrance for the Young," as it reads on every label. Back then, it was known as Violetas Russas, Russian Violets. And the lettering on the labels of the bottles was done in 24-karat gold leaf. Yet it wasn't expensive: about $2 for a five-ounce bottle. A Havana perfumer named Agustín Reyes created it in the 1920s, and it became an instant, permanent hit all across the island. (It wasn't only for babies' hair, either. Men used it as an aftershave and women sprinkled their bed linens with it.)

The initiation into Jubana femininity and identity officially began with Violetas. Later in life, Americanos would always comment on the "exotic" smell of my hair. Even now, if I need that Proustian madeleine-itude, I just reach for the bottle—of Violetas—and twist off the cap. I take a deep breath. My whole body relaxes. It's almost as good as Ativan, my pharmaceutical sedative of choice. Holly Golightly required a taxi ride to Tiffany to calm her nerves. She said something like, "Isn't [Tiffany's] marvelous? You feel as though nothing bad could ever happen to you here." That's how Agua de Violetas makes me feel (although I totally respect the Tiffany's mentality, too). Speaking of which, Mami loved *Breakfast at Tiffany's* so much that she nicknamed me Lulamae, Audrey Hepburn's character's "real" name. She still starts e-mails to me with *"Querida Lula Mae"* or *"Querida Luli,"* insisting that Truman Capote and Hollywood misspelled it, not she.

Cubans and Jubans are funny about names. Everyone gets nicknamed. On my birth certificate, it says "Beatriz Anders Benes." That is, of course, wrong. My given name is Rebeca Beatriz Anders y Benes. Rebeca is for my maternal grandfather Boris's mother. Mami Dearest wasn't crazy about it, but (1) Jews name kids after dead relatives and (2) she said it was the only thing her father had ever asked her to do for him.

Beatriz is for my father's maternal grandmother. Benes is Mami's maiden name. Though I personally like Rebeca and Beatriz, I've never been called either, not even derivatives of them. I think Gigi and its derivatives—Geeg, La Gig'—become my carbonated personality better anyway. Mami fell in love with Colette's *Gigi,* then Alan Jay Lerner and Vincente Minnelli's *Gigi,* and it stuck. Being nicknamed after a Gallic courtesan-in-training? *Pas de problème!* But in Cuba they called me Yiyi or La Yiya. One of Mami's 750 best friends was named Yiyi, and so that made it stick even more.

When I became a published writer twenty-five years later, I used Gigi Anders as my *Washington Post* byline. I once asked one of my Style section editors to let me use Rebeca B. Anders, just to see what it looked and felt like. She allowed me to do that one time, and never again, explaining that Gigi Anders was now my trademark, and perfect for a Style writer, a fun name evoking French poodles, rhinestone collars, bubble baths, and pink accessories, such as TaB. She was right.

The problem with Gigi versus Rebeca was school records, doctors' patient files, taxes, credit cards, health insurance, bank statements, social security and immigration documents, passports, and freelance check payments. Everything was constantly screwed up because nobody could ever figure out who I really was, legally speaking. As for me, it was all very confusing. In terms of identity, this was one massive matzo ball to tackle. If you're not even sure what your *name* is, where do you start to figure yourself out? Typically, Mami did not anticipate what difficulties her caprice would cause.

By 2001 I just couldn't take it anymore. So I went downtown to the cute little court building in Raleigh, North Carolina—I worked at the *Raleigh News & Observer* and, having moved to Raleigh from Washington, D.C., I thought everything appeared cute and little in comparison—and legally changed my name to Gigi Anders, thereby consolidating my personal, professional, and legal identities, and completely upsetting Mami. Legally changing your name is like getting a tattoo: You don't want to tell your parents beforehand. You have to present it to them as a fait accompli.

The following Thanksgiving up in suburban Germantown, Maryland, I was seated with my family (minus Papi, who was not up to it, and the fiancé, who had temporarily broken up with me in order to, among other things, avoid this very Juban social obligation) at my brother Big Al and sister-in-law Andrea's tarmac-size tiger maple French Country dining room table.

"I legally changed my name," I announced to no one in particular. Nobody was listening. You couldn't hear much above the din of Juban chewing and slurping. I cleared my throat.

"Hello!" I hollered. "Hello?" A herd of well-coiffed heads full of turkey, *gahntze tzimmes* (a big, sweet Jewish stew of brisket, carrots, prunes, dried apricots, and sweet potatoes), and *frijoles negros* (black beans) turned in my direction. "I wanted to let you all know that I legally changed my name."

"*¿Qué?*" Mami said, puzzled.

"Yep. I am now legally Gigi Anders. I went down to the Raleigh courthouse and—"

"Wow," said Big Al, aka Big Red, high-fiving me with the tip of his black bean–encrusted drumstick. "Pretty cool, Geeg!" The boy is a Little League coach who somehow has managed to turn himself into a physician.

Mami was still fixated on my latest name-dropping bombshell.

"Say again, joo went WHERE an' deed WHAT?"

"Courthouse," I said. "Legally changed my name."

"To WHAT?" You could tell Mami was getting worked up because she emphasized each word by stabbing the table with the lit end of her Kool. Observing this, Andrea, who's vigilant over her furniture and disallows smoking in her home, silently, mentally strangled her.

"To WHAT?" Mami repeated.

Now here's the thing: Mami has perfect hearing and a master's in psychiatric social work. She looks and acts like a young, redheaded Lana Turner. She's lived in this country and has spoken its language, sort of, for more than four decades. She's a worldly, accomplished, professional person with more outfits than poor dead Princess Diana ever dreamed of owning in her sadly abbreviated lifetime, and more shoes than Imelda Marcos (in fact, Papi's nickname for Mami is Imelda). Yet Mami makes me repeat and reexplicate everything I say as though she's a dim-witted, illiterate, recently arrived, monolingual refugee.

"What do you mean, 'to what?' " I told her. "To Gigi Anders. Have you ever once called me Rebeca? Rebby? Bex? Beca? Becky? Beatriz? Bea? Beckay? Triz?"

"Triz?" Dr. Big Red Al said. "What the F is that?" He recapped his gigantic jockstrap laugh, shaking his abnormally huge head of fire-red hair. Andrea languidly glanced at him with no expression, returning her gaze to the two smoldering spots darkening her imported tiger maple runway table.

"No, you have not," I continued, trying futilely to be logical and unsentimental—two adjectives rarely associated with Jubans. "I mostly did this because my taxes—"

Andrea was now up and alternately patting the table burns with a damp cloth and some orange oil.

"*My father,*" Mami tearfully commenced, verbally italicizing every word, "*my father never ask-ed me to do anytheengh for heem. De only theengh my father ever ask-ed me to do was to name joo after hees mother.*"

Satisfied with her ministrations, Andrea, the soul of Jewish-American tact and brown lipstick, correctly determined it was time to extricate herself and the babies, my niece, Lauren, and my nephew, Jack, from what smelled like a looming Juban shit storm. She feels lucky not to be Cuban at such moments, and really, who could blame her?

"Ooo-kay, it's time to say good night," Andrea said, dipping each baby straddled on her hips to kiss each guest.

"Well, you can call me Rebeca or Beatriz anytime you like," I told Mami, kissing Lauren and Jack. "It's still on my birth certificate."

Mami shot me The Look. It is not a Good Look. It's a withering combo platter of brutal disapproval, derision, outrage, insultedness, condescension, resentment, woundedness, and wild rage. It's beyond words.

Mami wouldn't speak to me for the entire 22.41 miles home, seething silence being one of her favorite punitive techniques. She just turned up Vivaldi's *Il cimento dell'Armonia e dell'-Invenzione* (Trial Between Harmony and Invention) to decibels heretofore unknown to civilization, lit up a Kool, and drove like a fucking NASCAR freak. She doesn't think of a car as a motor vehicle; for her, it's a weapon. Years ago I gave Mami a Cathy Guisewite key chain with a picture of a crazed Cathy behind a steering wheel. The caption read, "I OWN the road!" Mami took it to heart, and whenever she's in this state of mind, even hard-core maniac Capital Beltway drivers timidly yield to her white Honda SUV.

"I like to scare dem," she always says. "Sohkehrz."

Yiyita, my grandfather Boris called me. Born in poverty in czarist Russia, Zeide (Grandpa in Yiddish) Boris came to Cuba from Kraisk, a place so obscure that no map shows it, in Belarus, near the capital, Minsk. In 1941, during the German bombing, Minsk was pretty much decimated, and my grandfather's parents and siblings were killed, or so we believe. They didn't exactly keep strict shtetl records in those days.

Zeide Boris never spoke of his childhood. He said he didn't want to burden us with so many sad stories. Mami gleaned a few biographical details: Zeide Boris had never had toys back in Russia, so he improvised by playing with stray bullet shells and casings he found on the shtetl grounds. He had two brothers, and the parents, one of whom I'm named after, only had enough money to get one son out of Russia in order to not have to serve in the army. So they sent Boris, who at sixteen was within the then-draft age. My grandfather's skin was dark and swarthy, olive. His thick, straight hair was so shiny and black it was almost blue, like Hawaiians'. He was big, compact, dignified, and hardly ever smiled. He loved fine cigars and playing *dominós* and drinking a shot of good whiskey every night after work. Zeide Boris had a wise, compassionate, sad, faraway look behind his green semi-Asian eyes. Mongolian eyes. *Achinados,* Cubans call it. Chinese-ified. When I was a little child I once asked Mami why Zeide Boris had such unusual and striking features.

"Are joo keedeengh?" she answered, incredulous. "Back den een dos days everybody rap-ed everybody else! Pogroms, joo know? What ees a pogrom?"

"When they come and hurt poor Jews, like in *Fiddler on the Roof.*"

"Das right."

"And when they come and hurt poor black people, like in *To Kill a Mockingbird.*"

"Exactly! Ohldoh Gregory Peck was so beauteefool den, joo know. Ahteekohs Feench. I lohvee."

"And when they come and hurt poor women, like Dulcinea in *Man of La Mancha.*"

"Eet was johs rape, rape, rapeenghs," Mami said, "all throughout de heestohrees. Johs beeleeohns an' zeeleeohns of eh-sperms an' eggs just goheengh krehsee! Dat ees why der ees no such theengh as a 'pure' race. Pure, dat ees totally crap!"

"Ivory soap, that's pure."

"Not all de way. Ees only 99 44/100s percen' pure. I prefer Dove an' so should joo. Ivory's too harsh, joo know?"

"I don't feel clean when I use Dove. It doesn't lather up. I like that squeaky—"

"Dat ees so wrong! Squeaky ees really, really bad. Squeaky means dry, joo know? Dove ees better because eet moisturizes an' das what joo want every damn day."

"I just SAID that I didn't want that because—"

"An' also Dove ees a symbol of freedom an' peace," Mami continued, cutting me off. "Ivory ees what?"

"What do you mean, 'it's what?' What the fuck are you talking about?"

"*Ay, que bruta!*" Oh, what a dunce! "Ivory ees from de many, many, many dead elephan's een Ahfreeca dat dey steal for de eh-tusks. So many dead ones! Das really, really bad. Also eeleegahl."

Sigh.

Mami has normal skin and never uses an *après* bath body lotion. If I did that I'd desiccate. Mami must be blessed with low-maintenance, naturally soft skin. I, on the other hand, am so not. With the exception of the size, shape, and color of my nice but Helen Keller–blind green-golden eyes, I've had to work it and work it hard from head frizz to discolored toenail (from self-tanning overdose, hence the perpetual need for red polish) my whole life just to approximate nonhag normality.

I like to be fairly squeaky out of the shower, then I use industrial strength lotion (Lubriderm Advanced Therapy for Extra-Dry Skin or Johnson's Creamy Baby Oil, or, in the summer months, Lubriderm Daily UV Lotion with SPF 15—I'm delicate and I fry in the sun) from the neck down while I'm still damp, giving special emphasis to elbows (with the addition of Aquaphor Healing Ointment), butt (aforementioned Lube mixed with any good firming body lotion containing alpha hydroxy acid—you'd never want an unsmooth, saggy butt—plus Blisslab's Lemon + Sage Soft Oil

Spray Silky Soothing Skin Soak, which smells delicious and helps the first two ingredients to "slide"), and feet (regular Lube and several squirts of Johnson's Baby Oil, or Burt's Bees' Coconut Foot Creme with Vitamin E, followed immediately by a pair of thick white cotton socks). Then I sprinkle Johnson's Baby Powder with Aloe & Vitamin E Pure Cornstarch under my breasts and in between my legs to absorb any excess lubrication. Otherwise I'd literally stick to myself all day. (I repeat the baby powder routine at night before I go to bed because, at risk of sounding like a douche commercial, it makes me feel fresh.)

Unfortunately, this means that everything in my bathroom—the floor, especially—perpetually has a filmy white layer of *polvo*, powder, all over it. Which is why I'm a maniac about cleaning my bathroom at least once a week.

At any rate, my feet are so soft from having performed the daily moisturizing sacrament for decades, that some years ago my friend Sharyn began referring to them as "the pounded veal cutlets," which eventually got shortened to "the PVCs" or simply "the cutlets." A navel-pierced *Raleigh News & Observer* colleague once noted that my toes are so little and round, they look like grapes. My Vietnamese aesthetician, Christine, always says there's no point in charging me for a pedicure because there's nothing to do to my feet except paint the toenails. She means there's no yucky dry skin, calluses, or corns to fix, smooth, cut, exfoliate, and/or sandblast. The fiancé once remarked that my entire body is a cutlet because I take care of it. That was one of the nicest compliments I've ever gotten. What do American women do, just never moisturize? And who are these men's first wives who don't?

Suddenly bored with both the topic of the right body soap and the fact that I wasn't agreeing with her choice, Mami abruptly changed thematic lanes without signaling, signaling being for sohkehrz.

"De point ees," she declared, lighting a Kool and exhaling an elegant plume of mentholated smoke into my face, "everybody ees an eenterracial meex of everytheengh. Probably Heetlehr was part mulatto like hees demon eh-spawn, Feedehl. Dey both bought eento dat beegehst lie dat white ees right. Dey were de two beegehst, most self-hayteengh beegohts! Dat was der whole problem! Ees so ohbveeohs."

"*What* was their whole problem?" I asked, dodging her Kool line of fire.

"Dat dey couldn't accept een dehmselv-es, 'I AM MEEX-ED.' "

"Well," I told her, "then that means *we're* mixed."

"Das right," Mami said. "But de deefehrehnce ees, we lohvee!"

Zeide Boris met Dora Baicowitz in Matanzas, the lovely Cuban province just east of Havana known as the Athens of Cuba. They married in 1929.

Baba (our Yiddish version of Grandma) Dora was a buxom, attractive, fair-haired woman whose round nose, pale skin, and pendulous breasts my mother and I both inherited. By the time Baba Dora and Zeide Boris had finished having their children, in the thirties and early forties, there were some twelve thousand Jews in Cuba, out of a then-total population of seven million. (Today it's more like eight hundred Jews and twelve million Gentile Cubans.) The Benes kids were Jaime (blond, low-key, and easygoing, with his father's Asian eyes), Mami (redhead, headstrong, tempestuous, attractive, narcissistic), and Bernardo (redhead, brainy, prone to inappropriate and very loud opinionated emotional outbursts, visionary, narcissistic). They were officially Privileged People, high-ranking members of Havana's Jewish elite. Zeide Boris may not have been the wealthiest Jubano, but he was certainly among the top twenty wealthiest.

Mami was Zeide Boris's favorite. She was a serious child, didn't smile much, and watched everything around her very intensely. She took ballet classes (de rigueur for upper-class Cuban girls), loved music and pretending she could play the piano. When Tio Bernardo was born a year and four months later, he and Mami bonded like twins; they were close in age, appearance, and superior attitude.

Being a girl, Mami wasn't supposed to be good in math and science, so she wasn't. Besides, Zeide Boris never expected her to be the breadwinner once she got married. That's why, before tests and final exams, Zeide would always tell Bernardo, "Let your sister copy." This created a trend that spread. During one particularly tough exam, Bernardo uncharacteristically got some stuff wrong, and everybody else in the class got it wrong, too, because they'd all copied from him.

"De professor was so puzzl-ed," Mami recalls. "He was, like, 'Maybe I taught eet wrong?' "

Later in life, when my two less intellectually inclined American-born brothers would be challenged with writing assignments that I could have nailed in my sleep, Mami would get incensed if I refused to do their work for them, particularly the essay portions of their college and med school and business school applications.

From Mami's point of view while I was growing up, if you worked hard without cutting corners, asking for favors, using your feminine wiles or your ethnic minority-ness, you were a sohkehr. When I was a schoolgirl and Mami saw me poring over homework for hours, she'd get a stricken, pained look and say,

"*Ay mamita*, joor workeengh so *hard*." As if making an effort for more than two minutes was akin to working on a lesbian chain gang in the Mississippi Delta heat: an unattractive, back-breaking, deadening thing. Way too masculine a pursuit for a girlie gal. That quintessentially Cuban position is violently at odds with both the typical working-class illegal economic immigrant position, which is to work like a mule until you're dead, as well as classic Judaic values, which stipulate that talent is beautiful, that it's good to be smart—and that applies to girls, too.

When you're bicultural, oops, make that TRIcultural (I'm American, too), not every message you're given by your family and Juban community is synchronized or harmonious. Sometimes they switch into Cuban expectation gear, other times only the Jew 'tude will do, and at other times a nonethnic North American performance and persona is what's called for. Wouldn't *you* be a Messopotamia trying to integrate all that? It's one thing to be fluent in Spanglish, which I obviously am. But being triculturally fluent, now that's an art form unto itself.

Which self was I supposed to be, and when, exactly?

My family's apprehension and cluelessness about my aspiration to be a professional writer seem slightly odd, since Mami, who from birth bucked trends, has always been a professional. That was a huge big fact in the fifties, when most upper-class Cuban women didn't work. My mother certainly didn't have to. Her parents were filthy rich, and my father was doing great as the general practitioner co-owner of Centro Médico Nacional, his private hospital in Havana. But Mami wanted to work. Bored easily, she needed a lot of different kinds of stimulation, much more than running a household and raising a little baby girl could offer. She earned her master's degree in social work from La Universidad de la Habana, and took a position at the Ministerio de Bienestar Social, Cuba's social welfare department. When it comes to work, my civil servant mother has always been a driven, disciplined, competent, energetic, multi-award-winning achiever.

In Cuba, all her newlywed girlfriends were upset when she "abandon-ed" them for an office job. They even offered to pool their considerable financial resources to pay her salary so Mami would quit and go back to playing cards with them in the afternoon. But like me, Mami has always been on a different wavelength than the people around her (she also looks light-years better than her coworkers and makes it look effortless). The difference is that Baba Dora supported Mami's differentness, whereas my differentness has always been something that Mami *deals with*—what choice does she have?—but does not exactly celebrate.

"My mom joost to tell me how happy chee was dat I had a profession an' dat I work-ed," Mami says. "Chee said dat her generation deedn't have dat choice, an' dat chee would have lohv-ed to have had de opportuneety to do de same."

However, Zeide Boris's ideas about getting ahead in life—"Let your sister copy"—stuck with Ana and Bernardo. Mami was and remains deferential to Bernardo, whom she calls *"un genio,"* a genius. She has a blind spot when it comes to his *mishigas*. This has created friction, because I don't. That abrasive, macho, finger-snapping *"Tráeme un café AHORA"*—Bring me a coffee NOW—attitude compels me to verbally electrocute him. This outrages him. Tio Nano can be very hyper and domineering, especially around females. This subspecies of *Homo jubano* brings out the *Homo jubana americana ovaria* strap-ons in me every time. As for Zeide Boris, Mami likes to say that he was "a lion" when it came to defending his children. She remembers her father as being creative and hysterically funny. I remember a big, sad, elegant Russian bear of a guy who bought me all the *Archie* comic books I ever wanted.

Mami's impressions and recollections are her right, though

they do not always jibe with the facts. My former psychoanalyst, Marvin L. Adland, says that when it comes to memory and perception, what you *believe* is true is more important than what *is* true. That must be the case in my family.

Okay. I'm out of Mami and home in my beautiful new hand-painted, imported crib, canopied with a *mosquitero,* mosquito netting, to prevent errant flies from getting anywhere, God forbid, near *la niña,* the little girl. I've got a huge dresser with hand-painted knobs and baby bears in assorted pastel colors, bulging with adorable outfits. I've got two very wealthy, young, movie star–gorgeous parents. I've got a *táta,* a nanny, named Panchita; a *cocinera,* a cook, named Carmen; and a *criada,* a housekeeper, named Candita, who is Panchita's sister. I've got four grandparents—two dominant babas, two semiwhipped zeides—and one glamorous godmother living nearby, who all think I'm the cutest thing alive. I've got hundreds of aunts, uncles, cousins, friends, and parties. I have a zillion fantastic dolls, stuffed animals, books, and assorted toys imported from FAO Schwarz, and accoutrements from El Encanto, The Enchantment, Havana's version of Bergdorf Goodman. I've got my own air-conditioning in my bedroom. My hair is full of Violetas, my pierced ears sport tiny cultured pearls, my wrists are bejeweled with pearl and 18-karat rose gold ID bracelets, my pima cotton T-shirts, jumpers, and dresses are pinned with an *azabache,* a jet-black semiprecious stone said to ward off the evil eye, and I get endless sterilized bottles full of heavily sugared warm milk every single time I cry.

"*Mi heredera,*" Mami whispered to me. My heiress.

I gazed up adoringly at her beautiful freckled face framed in mosquito netting as she stroked my chubby pink cheek with her long, cool, pale, tapered, perfectly manicured fingertips. Her green irises were flecked with gold, sun rays that ringed her pupils like petals. The girl with sunflower, not kaleidoscope, eyes.

"¡Tafetán!" Mami murmured. "¡Vamos, mamita, di tafetán!" Taffeta! Come on, little mama, say taffeta!

Since I was clueless *and* toothless, this was clearly not gonna happen.

"Champán," Mami added. "Tafetán color champán." Champagne-colored taffeta.

It took about a year of hearing this bizarre mantra over and over before I was old enough to finally understand what the fuck my mother was talking about: the color and fabric of my wedding dress.

In the 1950s cha-cha-cha Cuba was the center of the Jubanite universe. Mami says that back then, Cuba had all the modern technology and conveniences of the United States, coupled with the Old World charm, grandeur, culture, and romance of Europe and Africa. It was simultaneously sophisticated and bohemian. Look at the black-and-white photographs from that era, my family's, anyway: Nobody looks unhappy, hungry, ugly. Our *comunidad Jubana* was *muy bonita*, beautifully dressed and beautiful.

Yet we were hardly *arrivistes*, historically speaking. Sephardic Jews were on the *Santa Maria* alongside el Almirante Cristobal Colón, Admiral Christopher Columbus, when they marched en masse out of the soft blue waves through the sugar-white sands of the Cuban beach in October 1492 to claim the island for España. The Jews were Spaniards called *conversos*, or Maranos, Jewish converts to Christianity. They were fleeing those outrageous anti-Semites Ferdinand and Isabella, and the king's close, personal, sicko *amigo*, Tomas de Torquemada, the Inquisition's twisted *coordinador*. One Marano, Luis de Torres, was Columbus's interpreter, being fluent in Hebrew, Spanish, Aramaic, and Arabic. De Torres is especially dear to my heart—and lungs—because he first observed and recorded tobacco smoking on the island. He wrote of seeing "many people, women as well as men, with a flaming stick of herb in their hands, taking in its aromatic smell from time to time."

Columbus found Cuba, which he called the pearl of the Antilles, the most beautiful place human eyes had ever seen. That's what he reported to his wacky Jew-hating patrons back home in Madrid. *La tierra mas fermosa* (it's really *hermosa*, but in old Spanish they didn't use H's) *que ojos humanos han visto*. The most beautiful land human eyes have ever seen.

What must my then-teenaged Eastern European and Russian shtetl-reared grandparents, aunts, and uncles have been feeling and thinking, pulling into Havana harbor? Were they apprehensive? Happy? Frightened? Dazzled? Lonesome? Relieved? Completely weirded out? All that fierce tropical beauty and shimmering heat, so far removed from the cold, bleak *Schindler's List* grayness of home. Here was a bird of paradise dreamland in blinding Technicolor, like Dorothy Gale falling asleep in black-and-white Kansas and awakening in colorful Oz. In Cuba the air was steamy and salty, and through it flew tiny *colibrís*, hummingbirds, and *cotorras*, parrots. There were palm trees and coconuts, plantains and *fruta bomba*, papayas. (In Cuban Spanish, the word *papaya* is slang for vagina, so we call the actual fruit *fruta bomba*. Which sounds good for vaginal slang, too, come to think of it.)

Maybe it was Eden, as dreamlike and foreign as Hieronymus Bosch's *Garden of Earthly Delights*.

Unreal. That's what my ancestors thought. *We're on a different planet now.*

In my parents' and grandparents' time—roughly from 1920 through 1961—Cuba was, for the most part, religiously and racially tolerant. My family never spoke of experiencing much anti-Semitism. My father's parents, the less well-to-do Leon and Zelda Andursky, hailed from Poland. In Cuba, the slang word for Jews of any nationality was *polaco,* Polack. *Oye polaco, ¿qué pasa?* It was a term of endearment among ourselves, but definitely off-limits for non-Jews. (It's like blacks calling one another nigger. They can, you can't.) Gentiles would call us *los hebreos,* the Hebrews, never *los judios.* My Andursky *abuelos* never changed their surname, but Papi did, to Anders, when he began practicing medicine. At the Centro Médico Nacional, he'd overheard hushed comments about "Cuba is for Cubans" and *"el médico polaco."* Papi didn't like it. I love Anders. I think it's a great name. It sounds like Switzerland. In German, the word actually means "different" or "otherwise." Plus Anders is the perfect neutral foil for Gigi. It tempers the bubble-bath/poodle/rhinestone connotations, not to mention it beats the shit out of Gigi Andursky as a byline.

Meanwhile, Hollywood movie stars and assorted celebs and politicos were flocking to our island to play: Ava Gardner, Frank Sinatra, Groucho Marx, Dorothy Lamour, Maurice Chevalier, Eartha Kitt, George Raft, Edith Piaf, Cab Calloway, Dorothy Dandridge, Tony Martin, Jennifer Jones, David Selznick, Marlon Brando, Pablo Picasso, and Jesse Owens (who raced against—and beat—a horse). Hemingway was in his *Old Man and the Sea* prime, Winston Churchill couldn't get enough of our posh casinos, country clubs, or cigars. And the Mafia, well, you saw *Godfa-*

ther II, right? The $14 million Hotel Riviera, for example, was financed mostly by the Cuban government for Meyer Lansky. (They had a floor show in the Copa Room headlined by Ginger Rogers. Lansky noted that "Rogers can wiggle her ass, but she can't sing a goddamn note.")

Papi said Cuba was a corrupt place under the crooked, ruthless dictator General Fulgencio Batista's rule, and corrupt before him, but it was an alluring, sexy, prosperous, lush, advanced, beguiling, laissez-faire kind of corrupt. You know, fun corrupt. American musical acts went there to play all the time. That's why Cuban Americans of my parents' generation think it's hilarious that the 1997 *Buena Vista Social Club* CD was such a hit in this country, as if *los Yanquis* were just discovering the sinuous beauty and earthy soulfulness of our native music.

I asked my mom's best friend, Eliana, what she thought of the record, and she said, "Dat cheet? Dat was, like, *música del campo* [country music]. Tacky. Een Cuba I leesehn-ed to Nahpkeen Kohl." (She always called him Napkin.)

Mami used to listen to "The Christmas Song" crooner, too, until she met Nat King Cole one day at a baseball game in Havana. She ran over to him, confident and full of teenaged life, all freckles and bosoms and red lipstick, and breathlessly asked for an autograph.

Cole slowly sized her up from his seat, paused, frowned, and condescendingly said, "No."

Mami's never forgiven a no from ANYbody.

"Wow," she sputtered. "Joo are really an ASShole."

She's hated Nat King Cole's guts ever since.

Following the earlobe trauma, the other key thing that happens in early Jubanahood is that by age two or so, people stop feeding you just plain old *leche* and start with the *café con leche* at breakfast. You know how they say that one drop of black blood in a glass of white milk makes you racially a chocolate milk? Well, one drop of espresso in a glass of heavily sugared hot *leche* makes you wired. It permanently alters your already nervous system and turns you into a caffeine and sugar addict with an attitude. (So the next time you ask yourself, "What is it with these fucking Cubans?"—remember the *café con leche*.)

The trick about *café con leche* is that it's initially soothing but ultimately stimulating. Soon after I started drinking it every morning I began speaking in whole sentences. I haven't been able to stop myself ever since. When my parents had parties, which was constantly, I'd sit on the living room's cool Spanish tile floor in a pair of black ballet tights and nothing else (except the jewelry, of course) and explain the meaning and causes of thunder and lightning. Cuba's known to have hurricanes every now and then, so I knew this would be a big hit for my audience.

"En resumidas cuentas, todo esto tiene que ver con fuerzas negativas y fuerzas positivas," I commenced. *"Como la vida misma."* The bottom line is that it all comes down to negative forces and positive forces. Like life itself.

The well-heeled *invitados* were *muy* impressed. They approvingly sipped their daiquiris, Bacardi Cuba Libres, and Manischewitz.

"She looks so much like David," people always told Mami, "you could dress her in a white lab coat and send her to the hospital. No one would know the difference."

"I don' agree," Mami always replied, even though she knew it was true. *"La niña* ees very Benes. Her hair ees rayt, like mine an' Bernardo's."

No. My hair was and really is auburn, with golden-red highlights. But my mother has always considered redheads as well as her side of the family infinitely superior to everyone else. Certainly the Beneses were richer and more refined than the Andurskys, but none of my grandparents had ever had more than an elementary school education. Papi's parents were comfortable, middle-class. They'd emigrated as newlyweds in 1927 or '28 from Zalnik, Poland. It's an obscure village like Kraisk, that no map shows. Papi was born shortly after they'd arrived, in 1929. Zeide Leon owned Cuban American Textiles, in La Habana Vieja, Old Havana, at #557, Calle Compostela. He and my grandmother, Grandma Zelda—"Baba Zoila" in Yiddish and Spanish—sold upholstery fabric for sofas, chairs, and curtains. They also sold fabric for men's suits and women's dresses. The store had no a.c., just two huge standing fans. Mami says in the summer the heat in there was unbearable.

Zeide Leon was short and strong, wiry, well-built. He smelled like steam, from his freshly pressed white short-sleeved shirts, and Aqua Velva. On Sunday afternoons we'd stroll through El Jardín Botánico in Miramar or El Parque Central in La Habana Vieja. I'd pick flowers to bring home to my mother while Zeide Leon kept an eye out for stray cats. When he'd see one he'd pummel it with stones and laugh maniacally. I'd scream and run under Baba Zoila's full skirts to hide, which only made Zeide laugh harder.

Baba Zoila was bossy, critical, and dyed her beehived hair a matte black-brown. She wore heavy red lipstick, covered her furniture in plastic sheeting, and made a lot of boiled chicken and Jell-O. She doted on my father and held him responsible for Tio Julio, Papi's brother who was five years younger. Baba Zoila held Mami in medium esteem at best. She thought Mami was self-centered and spoiled rotten. On our Sunday outings, Baba Zoila would tell me, "Your Papi is the most handsome, smart, giving, wonderful person in the world! The best son, brother, doctor, father, and husband! That's why he's named after a king. You know King David?"

"Uh-huh. What about my Mami?" I'd ask, shaking the dirt off the roots of my filched hibiscus or jasmine or frangipani or gardenias.

"Your Mami?" she'd reply. "She's . . . she's good, too."

Good? She was way more than *good.* She may not have been a biblical monarch but she was a goddamn goddess. And the fact that you didn't like it didn't change it; even I knew that. That's why I thought it was mighty big of Mami to give me the floor, if only fleetingly, at her sophisticated soirées. Here was a deity who could share the spotlight sometimes. I could never compete with her beauty, grace, and charisma, of course, or make Papi pay attention to me, but I could be funny, smart, charming, and *una pícara,* a cheeky girl.

So I pressed on with my thunder and my lightning.

"¡Y entonces hay un choque!" I said, clapping my tiny hands together once very loudly to illustrate the fury of *el choque,* the shock, as the two forces collide. That always gets their attention.

"Calor y frío," I continued. Hot and cold. I'd squeeze my eyes shut, turn my head away from the impact, and slam my palms together again. *"¡Ay, que escándalo!"* Oh, what a scandal!

Exhausted from my educational performance art, I'd drop backward and faint.

"Coño, super-mona, pero demasiado café con leche, tú," was my audience's unanimous verdict. Shit, she's super-cute but too much *café con leche,* man.

"A ella le gusta," said Mami, shrugging. She likes it.

Meanwhile, Tio Jaime was tickling the soles of my feet to revive me. Screaming with laughter, I tried kicking and kicking him away. He'd grab a black stockinged foot and drag me around as I shook my hips and swayed my abnormally large head (threatening to crack it open like a coconut on that hard Spanish tile) as I crooned the Elvis song I'd heard many times on Carmen's kitchen radio (she called him El Rey): " 'Baby, I ain't askin' much of you/Just a big-a big-a hunk o' love will do/If you'd give me just one sweet kiss, no no no no no no no . . .' "

"Y bueno, de todas maneras," Mami concluded, surveying me wiggling down on the floor from the chic elevation of her black peau de soie Dior open-toed slingback stilettos as she delicately dabbed the side of her shapely mouth with the expert tip of her pinky to assure her red lipstick wasn't wandering, *"la niña salió así."*

And anyway, the girl was born like this.

Mami's bête noire may be Napkin Cole. Papi's is Herbert L. Matthews. "Herbert 'El Cabrón [The Bastard]' Matthews," my father calls him. "It's all his fault. Yep. He was a dick."

Before I was born, in February 1957, the *New York Times* ran a three-part series written by Herbert L. Matthews, who'd gone to see Castro in his hideout in the Sierra Maestra, the rugged mountain range in southeast Cuba. As author Tad Szulc wrote, at age fifty-seven, Matthews was a highly respected, seasoned pro, a member of the paper's editorial board, who specialized in overseas reportage. Matthews was an elite: intellectual and reticent but fundamentally quixotic. Matthews's wildly sympathetic portrait of Castro and his followers turned Castro into a myth of goodness, guts, and social justice. The wonderfully written stories vastly influenced the planet's attitude toward that myth. After all, it's the *New York Times,* for God's sake.

"[Señor Castro] is a hero of the Cuban youth," Matthews wrote. "This was quite a man—a powerful six-footer, olive-skinned, full-faced, with a straggly beard. He was dressed in an olive gray fatigue uniform and carried a rifle with a telescopic sight, of which he was very proud . . . [Señor Castro is] an educated, dedicated fanatic, a man of ideals, of courage and of remarkable qualities of leadership . . . [who speaks with] extraordinary eloquence . . . He has strong ideas of liberty, democracy, social justice, the need to restore the Constitution, to hold elections . . . he is now invincible . . . invulnerable.

"The personality of the man is overpowering. It was easy to see that his men adored him and also to see why he has caught the imagination of the youth of Cuba all over the island . . . the best elements in Cuban life—the unspoiled youth, the honest business man, the politician of integrity, the patriotic Army officer—are getting together to assume power . . . they are giving their lives for an ideal and for their hopes of a clean, democratic . . . and therefore anti-Communist Cuba."

"*Coño*," Papi says, "*qué bruto ese Matthews. Qué ingenuo. Eso nos jodió.*" Dammit, what a dunce, that Matthews. What a naïf. That fucked us.

Well, if the august Herbert "El Cabrón" Matthews, who came to Cuba all the way from the *New York Times*, was mesmerized and taken in, then so were we. Because during that period, and much to Zeide Boris's horror, my very own mother and Uncle Bernardo were publicly going around with other young people clutching anti-Batista placards. Bernardo even got arrested once for organizing an anti-Batista protest rally at a Havana Sugar Kings baseball game in Havana's El Gran Estadio del Cerro. Naïve or not, Matthews's trilogy had provoked a virtually universal groundswell of support for Fidel Castro.

"What's *wrong* with you?" Zeide admonished Bernardo after a friend got him released from the slammer. "You'd have to be crazy to get yourself killed over politics."

Zeide Boris knew firsthand what communism looked and

smelled like. And sounded like: As he often said, "When you hear a bark, it must be a dog."

"Zeide had leev-ed through eet een Russia," Mami says. "He recogniz-ed Feedehl was joozeengh de same tacteecs. But Bernardo an' I were totally doop-ed."

Because Bernardo was totally committed to Castro, he and my grandfather got into terrible arguments. Tío Jaime, Mami says, was "very cool" about the coming revolution. Mami wasn't as passionate as Bernardo, but she wasn't on the fence like Jaime, and she was profoundly influenced by Bernardo's politics. For example, a law school classmate of Bernardo's, Osmel Francis de los Reyes, was also a virulent anti-Batista Cuban. But unlike my mother and Tío Bernardo, Osmel was being hunted by Batista's men as a traitor for his outspoken candor. So, like Anne Frank hiding in the secret annex with her family, Osmel holed up at my parents' apartment for six months in 1957 while Mami was pregnant with me. Mami says the mailman thought Osmel was Mami's lover because he'd stop by with the day's mail and see the two of them in their bathrobes, drinking their *cafés con leche* at the breakfast table. Mami and Bernardo eventually got Osmel asylum at the embassy of Brazil. Osmel moved to Brazil and lived there for years before returning to Cuba, a broken man with a bad drinking problem. I asked Mami why in the world she would take such a risk, especially while being pregnant, and if Osmel was Bernardo's friend, why didn't he go hide at Bernardo's damn house?

"Well, we were johng, joo know? We thought Feedehl Castro was de answer to all our problems."

The problem Mami had soon after Castro took power was no freedom of speech. At work she'd criticize Castro to some of her colleagues and "Deyd look at me to choht up. And I deedn't like to choht up. I had to go to de bathroom weeth my friends to talk. Joo become paranoid weeth good reason."

And reckless for no reason but thrills. Two days after the revolution in 1959, my mother and Bernardo insisted on getting in the car with Papi, who didn't want to go, and Tía Ricky, Bernardo's wife, who was pregnant with Joel, her first child, to go check out the scene downtown. They parked in front of the Hotel Nacional and, except for Papi, began getting out. Papi yelled, "Don't get out of the car!" but the others, as usual, didn't heed his warnings. Suddenly there was machine gun fire. Batista holdouts were firing down from the hotel's rooftop. Papi shouted, "Lie down!" and ran out of the car to protect Tía Ricky by covering her with his body. Nobody got shot or too hurt. The quartet was incredibly lucky and, except for Papi, incredibly stupid.

"Ees true," Mami says. "We *were* stoopeed to go. But joor johng an' joo don' theenk. We were fool of energy an' wanted to see what was goheengh on. Papi deedn't want to go an' we forc-ed heem. Papi *nació viejito.*"

Papi was born a little old man.

Another girl, a sweet little redhead, was born to Mami on June 6, 1960, a year and a half after the hypocritical Castro & Co. had overthrown the hypocritical Batista & Co. My new sister's name was Cecilia. Cecilia lived for only two days, dying of a congenital heart problem. Mami had had German measles during the pregnancy. Losing Ceci was the first major blow to my mother's otherwise charmed life, and she never recovered from it. She left the maternity ward like a ghost and locked herself in her bedroom for a week or a month or a year—it was all the same to me—and refused to talk or eat. All Mami wanted to do was sleep. Zeide Boris had always told her there'd be plenty of time for sleep later on (after death), but Mami slept and slept. I tried turning the faceted cut-crystal doorknob of the door to her bedroom, but it wouldn't open. Sometimes Mami's gorgeous blond friend Anitica came over; she was the only person allowed inside the sanctum. I pressed my ear up against the door lots of times but I never heard anything. Perhaps there was nothing to say.

My parents have almost never spoken of Cecilia, a person whose memory instantly turned taboo. There was no funeral. On the anniversary of her death, June 8, my parents do not light a yartzeit candle, as they do for other dead family members we honor and remember. I myself was able to fully remember the experience of Cecilia only when I got into psychoanalysis three decades later. My brain's hard drive finally recovered the buried file and it came back up intact, crystal doorknob and all. I told Mami about it. She was horrified. She was mad. She couldn't understand how I could possibly remember something that happened when I was so tiny. Mami decided she was suddenly ambivalent about Gramps (my nickname for my psychoanalyst, Dr. Adland), whom she'd originally liked. She accused me of having "too much memory." I said there's no such thing.

In 1970, ten years after Cecilia's death, in another country, Simon & Garfunkel's *Bridge over Troubled Water* LP came out. I was still just a little kid, but I'd play the Latin-flavored song "Cecilia" over and over, just to hear my sister's name: "Cecilia, you're breaking my heart . . . Oh, Cecilia, I'm down on my knees/I'm begging you please to come home/Come on home."

I loved how the singers drew out the second syllable, that long, lovely "ee" sound, Ceciiilia. Even now, whenever I miss my Ceci, or the idea of a sister whom I'll never have, I play the song. I know it's about a prostitute but it comforts me all the same. If I ever have a daughter, Cecilia will be her name.

• • •

On my second birthday, in December 1959, Mami went out to buy party supplies, leaving me home alone with our cook, Carmen. (Panchita and Candita had the day off.) Carmen made me the typical Cuban child's *merienda*, snack, of a *café con leche* and a *galleta Cubana con queso crema y guayaba*, a Cuban cracker with cream cheese and a slice of guava paste. She turned on the radio. This time there was no Elvis, El Rey; it was an angry man's voice yammering and yawping in Spanish.

"El Caballo," Carmen explained, with a surrendering, dreamy sigh. The Stallion, Castro's nom de guerre. She dabbed my cream cheese–covered lips with one of the imported embroidered white linen napkins my parents had received as a wedding gift.

"*¿Cuál caballo?*" I asked, licking the sticky *guayaba* off my pearl and rose gold ID bracelet. Which horse? "*Caballitos no hablan.*" Little horses don't talk.

Carmen laughed, quietly and knowingly.

"*Vamos, Yiyi,*" she said. "*Tenemos que vestirte para la fiesta. Tengo una sorpresa para ti.*" Let's go, Gigi. We have to get you dressed for the party. I have a surprise for you.

When Mami came home and saw me standing on her bed, modeling my new birthday outfit complete with all the requisite accessories, she screamed (in a bad way). Carmen, whose secret militiaman lover was a Castro convert, had dressed me in a baby guerrilla fighter uniform. Olive-gray camouflage shirt with matching cargo pants, tiny sandals, and a black Ché Guevara–style beret. (Sounds like a hip Gap Kids ad, doesn't it?) Oh, and a toy replica of Castro's beloved rifle with a telescopic sight, the same kind Herbert L. Matthews had described Fidel carrying and being so proud of.

"*¡Mi'ja no es una guerrillerita!*" Mami cried, yanking the plastic rifle out of my hands and throwing it at Carmen's head. My daughter is no baby guerrilla! *"¡Atrevida! ¡Ahora mismo te vas de mi casa, traidora, o te pongo de patas en la calle!"* You insolent woman! You leave my house right this second, traitor, or I'll kick you to the curb!

"Mi novio es miliciano," Carmen said tauntingly. My boyfriend is a member of (Castro's) militia. It was a clear threat.

"¡Pues te vas con tu novio el miliciano a la calle!" Then you go with your boyfriend the militiaman to the street! Mami maintained her outward bravado, but she instinctively knew this incident was a prologue of dark things to come.

"Á mi no m'importa un carajo," Carmen replied, indifferently. I don't give a fuck. She picked up the toy rifle and pointed it at us. *"Ustedes son los traidores. Ustedes son los que se van a arrepentir."* You guys are the traitors. You guys are the ones who'll regret it.

Years later, Mami would admit, as only she could, "What really freak-ed me out was dat joo look-ed so cute. But I couldn't, like, get behind eet."

In October 1960 Castro's new revolutionary government confiscated Zeide Boris's Camisetas Perro, Zeide Leon's Cuban American Textiles, and Papi's Centro Médico Nacional. All three men's assets were frozen. My father was still in his twenties, but my grandfathers and grandmothers, whose adopted Eden was sinking, were already well into their late fifties. The same fate befell thousands of others, as well as the owners of companies such as Bacardi, Colgate-Palmolive, Coca-Cola, Pepsi Cola, and ITT. Smelly guerrilla thugs barely old enough to shave arrived unannounced at our beautiful apartment. They had machine guns strapped across their chests. They took inventory of all our belongings and told us we could keep only whatever could fit into two or three suitcases apiece. In passing, they mentioned that they really liked our imported linens and my tiny pearl earrings and golden baby bracelets. They added that our spacious apartment would make an ideal romantic hideaway for El Caballo and his many lady friends.

April 9, 1961: A bomb explodes in El Encanto. Another bomb explodes near the Pepsi Cola factory.

A newspaper ad from that period refers to El Encanto's five floors and sixty-five departments. "While in Cuba, do not fail to visit El Encanto." (Mami says El Encanto had seven floors. Let's go with her version. It's not worth the aggravation to contradict her with "fact.") Mami coped, sort of, with bomb number one. Denial's always good. "Probably some twelve-jear-ol' gohreelah jerk was een der trying to choh off or sometheengh. Stoopeed fohkehr. Dees *guajiros* [peasants] can't appreciate anytheengh goo'. I johs pray eet wasn't een de shoe departmen'."

April 13, 1961: Another explosion at El Encanto reduces the seven-story building to dust and rubble. Just powder.

When her beloved Encanto was no more, Mami knew we were in real trouble. You just can't stay in a place when your favorite luxury department store isn't there anymore because two rebel bombs blew it up; your father, father-in-law, and husband have become suddenly unemployed and bankrupt; one of your baby daughters is dead; and the other one's dressed to kill, not marry.

CHAPTER TWO

Cubans in Snowflakes and
Wahndehr Brayt

*P*oor little Cuba. Everybody wants a piece of magical, tiny you. Beautiful hand-tinted picture-postcard place. Ripe and sweet, juicy and soulful. The Cuba my parents knew is gone. But even now, outsiders keep picking at the carcass. The Mafia has been replaced by European tourists who go there on vacation to have sex with hot, soft minors because the hot, soft minors need cold, hard cash to survive. That's a perverse fact for people like my parents, who were considered deserters and called *gusanos.*

Worms.

These people were forced out with only the suitcases they could carry.

I also got to keep the little red tricycle that Mami thought would get us all killed.

Mami had asked her friend Georgina to take care of our apartment while we were "away," and handed her the keys. I was two years old. Nisia, my godmother, drove us to the airport. The tricycle and stuffed lamb were in the car with our luggage.

Hasta luego to my *tata*, Panchita, who lovingly took care of and played with me, who fastened my pima cotton diapers with 18-karat gold pins and who mashed up black beans for me before I had teeth. *Adiós* to my *criada*, Candita, who carefully washed and polished the cool Spanish tile floor and ironed Papi's linen *guayaberas* and the hand-embroidered linen tablecloths and matching napkins. *Adiós y hasta nunca, puta* Carmen. Good-bye and good riddance, Carmen, you bitch-whore.

We left our spacious apartment on Calle Ocho in Miramar, the elegant northwest suburb of La Habana, for the ten-mile drive to José Martí airport. It was Tuesday, November 15, 1960. When we arrived there, I hoisted my lamb, tucked it under my arm, got on my tricycle, and pedaled furiously all the way to the airplane on the tarmac. The armed guerrilla at the foot of the steps told my mother that I couldn't take the tricycle on board. She nervously asked me in English—presumably so we wouldn't be understood by the guerrilla—to please get off the trike and let the guard have it.

"No!" I screamed. "This is MINE!"

"I'll get joo another one," she whispered, "a better, preettier one."

"Nooo!" My mother tried to pull me off the seat, but I wouldn't let go. "This is MY bike!" I roared, more to the guard than Mami. "It's my birthday bike! I'm not leaving it here! I'll kick that man if he tries!"

Miraculously, we were not shot on the spot.

"Okay, okay, *adelante*," the guerrilla finally said, taking a couple of steps backward and shaking his head, arms bent and palms facing out, as if to say, "I surrender." My mother carried my tricycle onto the plane. I carried the lamb.

Nisia stayed in Cuba by choice and is still there. She holds a high-ranking government position in the arts. I have a photograph of that last day in Cuba, my most cherished object. I carry it with me all the time in my Filofax: Holding me in her warm, full arms is a young and glamorous Nisia, her cheek and mine pressed against each other. We have never seen each other again.

In Cuba we were beautiful and affluent. In the United States we were beautiful and broke. Like so many other first-wave Cuban exiles, we had lost it all. When we arrived in the United States, we were sure this was a temporary situation. I always say "we" because although I was just a toddler, I felt like the third spouse in my parents' marriage. They were young, in their early twenties, and never talked down to me as though I were an idiot. I was treated, for the most part, like a short adult. So instead of teaching me normal nursery rhymes like whatever the Spanish-language equivalent would be of "The Itsy-Bitsy Spider," my mom taught me the one about how many long needles and pins I would stick in Fidel's body when he was dead and bloated, stinking on his back in the tropical sun. A subhuman pincushion with an ugly, pornographic beard. That's actually how I learned to count. The lyrics went, *"¿Cuántas agujas vamos a meter en Fidel, m'ija?"* How many needles will we stick into Fidel, my daughter? *"Vamos a meter una, dos, tres . . ."*

I think Mami made it up. Children of the wronged and the imaginative get inculcated early. I don't know how you accept the loss of your beloved home and country and everything and everyone you knew and loved in your life. Maybe you have to imagine that the loss isn't a permanent fact; that it may be real, but not *really* real. Displaced just as my four Ashkenazi grandparents had been a generation before, we became wandering Cuban Jews. Jubanos. My grandparents had lost their parents to pogroms or to the Holocaust. Had they remained in Europe and Russia during Hitler's rise, who knows what might have become of them?

One of my maternal grandmother's brothers, Zalman (the Hebrew transliteration of Solomon) didn't leave the shtetl in Lithuania in time; soon he was rounded up and forced into a cattle car and sent to a concentration camp until the end of the war. He survived. He immigrated to Cuba. He called me *la princesa*, the princess, and when he held me on his lap and stroked my face, I thought I could see little green numbers on his wrist. Tío Salmen (Uncle Zalman in Spanish) had never celebrated a birthday until he came to Cuba as a young man. Behind his deep-set pale watery eyes lay a mystery of an unspeakably sad past.

How much reality can you stand? How much hurt? Where is the grace, humor, and strength to pull you through? How do you let go of everything you love?

Sometimes you have no choice. And so, in America, we started over again.

Most of our relatives stayed in Miami, others went to New York and New Jersey, still others to North Carolina to work in the textile mills. My parents were recruited by officials from St. Elizabeths mental hospital in Southeast Washington, D.C., who had gone to Miami in search of medical professionals among the refugees.

Mami had one black dress, one string of pearls, one pair of pearl earrings, and one pair of black high heels, and she wore that outfit to every job interview she had. It was so cold when we arrived that November that we had to go buy "coats." Nobody in Cuba ever owned a "coat." There was no need. Baba Dora, my maternal grandmother, had given Mami her mink stole *para el frio allá,* for the cold up there in the North, and Mami got tired of having to sneak around wearing it, feeling totally out of place in a slummy neighborhood in Southeast Washington. A Jubana anachronism. So we went downtown to Hecht's department store. They had a bakery there, and she'd buy me an elephant ear to quiet me while she shopped. Mami was always shivering and crying and cursing the weather because she was so homesick and she hated the cold and the snow. We'd warm up in one store, dart out into the freezing street, then run into another store to warm up again.

Mami went to work after about a month, leaving me with a very sweet fellow Cuban refugee, a tall, skinny black woman named Amelia Gutierrez whom we'd met in Miami. We loved her and she lived with us for a year. After my bath Amelia would slather me with Alpha Keri body lotion until I was sticky (which I couldn't stand, which is why I'm so into the baby powder cornstarch attitude now). She'd make us lunch, varying typical Hispanic meals like fried eggs on white rice and fried ripe plantains, with typical American ones like pasta with butter and salt. Amelia was a wonderful seamstress. On Saturdays, after our mailman stopped in for his weekly espresso (he'd become addicted after the first time Mami invited him in for a *cafesito* break), Amelia, Mami, and I would go to *el tehng cehng*, the 5&10-cent store, to buy fabric to make dresses for the three of us. Mami and I were very sad when Amelia decided to go to New York to live with a cousin.

On my fourth birthday, I got a new nanny, a Peruvian kindergarten teacher whose name no one can rcmcmber now because she stayed so briefly. This lady was also really nice and talented. She knitted and crocheted me the most delicious winter party dress, my first. It was a little girl's bespoke dream, like wearing spun cotton candy, all pale pink angora, with tiny seed pearls hand-stitched into the bodice. Anastasia Romanov herself couldn't have had a better one.

Because Zeide Boris was Russian-born, I cultivated a fantasy that I *was* Anastasia, cruelly banished from my homeland, wandering the earth, misunderstood and ravaged, English a little rusty, a haunted princess. Yet I still bore that unmistakable

Russian refinement, wreathed in my sole pathetic surviving possessions: tiny pearl earrings and a custom-made pink party dress. I was *poignant*. Of course, *my* royal parents had made it out of the Revolution with me, and that certainly was a consolation. But everything we knew and loved—our riches, our whole old world, the lovely cushioned and cosseted way of life—had been wrenched from us by disgusting, illiterate revolutionary brutes. As our real-life family priest Máximo says, it's much worse to have had everything and lose it than to have never had anything at all. Máximo is a fellow Cuban exile whose fabulously wealthy family left Cuba almost exactly when we did. He is a Catholic priest in Washington, D.C., with a weakness for Neiman Marcus. Ergo, Mami Dearest considers him her brother. He gave her a medallion of St. Joseph of Cupertino (patron saint of air travelers, who levitated while praying), which she keeps in the coin compartment of her Coach wallet. Mami suffers from acute fear of flying—triggered when we flew out of Cuba forever—and unfortunately I inherited that phobia in my late twenties and still have it.

Though I got to keep my lamb and tricycle and we flew safely over an ocean and landed just fine, somewhere a trauma clicked and got locked into place. Maybe it was because someone at the Miami airport managed to steal my beloved little red birthday tricycle while we were being interviewed by immigration officers. Whatever it is, ever since I can remember, I've had the same two recurring nightmares. One is that someone is stealing something from me and I see them doing it and I accuse them and they always get away with the theft, no matter how much I yell and scream about the injustice. I awake hoarse and frustrated and humiliated. The second one—this one scares the bejesus out of me—is that I'm in *cinquecento* Venice, sightseeing. The permeating light is that ancient, thin, transparent yellow light seen in oil paintings from that period. I'm with an older man on a suspension bridge, about a mile above the city, only there are no towers supporting the bridge. It's almost like a bridge made of vines, like from Tarzan movies. The bridge sways and feels very flimsy underfoot. The man and I are crossing it to reach the other side of intricate canals, moldering palaces, and secret Leonardo grottos. The dirty water below is silent but portentous. It's a hazy, murky green-gray, like a lagoon or a swamp, but deep enough to drown in. A black crow shoots across the sky like an arrow, and my companion turns abruptly to look at it, agitating the bridge. I grasp the rails and they're rope. I look down and I'm losing my balance and already feel asphyxiated because I know I'm going to fall into fathoms of mysterious water. I wake up to not drown.

But when I was a fearless teenager, I just loved traveling by plane. The more takeoffs and landings, the better. I was sixteen when my parents and I flew to Mexico for a holiday. We stayed with friends in Mexico City. Our hostess, the gorgeous bohemian wife, Nedda, was one of my mother's best friends in Cuba. With her thick blond hair (the kind that could break a hairbrush), green-blue eyes, tawny skin, and perfect figure, Nedda was considered possibly the most beautiful girl on the entire island in the 1950s. Papi dated Nedda when Mami broke up with him during their courtship. Early on in their relationship, my father had come to pick my mother up for a date one night. Mami was seventeen, Papi was twenty-one. Remember, he was a relatively poor boy going out with one of the wealthiest girls in the tiny Cuban Jewish community. When he arrived at her grand house, Papi told Mami that they'd have to take *la guagua*, the bus, because his car was broken. Mami said okay. But then Papi said, "No, I have the car. I was just testing you to see if you'd accept riding *la guagua*."

She dumped him on the spot.

Papi was devastated. He lost a lot of weight. But he eventually got back at Mami by dating her girlfriend: Nedda. When Mami saw them out together she had a sudden change of heart. Papi made her beg for a looong time before he would deign to take her back. Mami called Zeide Leon and asked him to help her. Zeide Leon encouraged Papi to go see Mami. Mami convinced Papi she'd suffered from temporary insanity. Papi said, "Okay. Let's give it a try." That would be the last time Mami would ever beg Papi for anything.

Nedda's suave Mexican husband, Enrique, was Old World glamorous, like Oscar de la Renta but not bald. Their house smelled of her signature scent, Calandre, and had so much original artwork that paintings were even hung on curtains. We ate roasted *cabrito* (goat) and steamed cactus. Mami admired Nedda's collection of black clay pottery, which Nedda explained was a specialty made in Oaxaca. Naturally, Mami had to have some—partly for herself, a compulsive tchotchkes hoarder, and partly to resell for profit at St. Elizabeths to staff and patients alike, whoever could pay—and insisted we fly down there, which would take less than an hour. Papi, of course, didn't want to go; he's never wanted to go anywhere or do anything. It was a miracle he came with us to Mexico in the first place. He hates having "new experiences" and meeting "new people."

But *comme d'habitude,* as always, Mami got her way. The following overcast afternoon the three of us were en route to Oaxaca. Everything was fine for about ten minutes. Then the sky went dark and rainy, and we hit some serious turbulence.

MAMI (hysterical): *¡Ay, Dios mio! ¡Ay, Dios mio!*

PAPI (semiconcerned): *¿Qué pasa?*

MAMI (annoyed): What do joo mean, *¿qué pasa?*

PAPI (getting nervous about bleeding to death in flight from my mother's sixteen-inch painted talons digging into his exposed arm skin): *Coño, cálmate. Esto no es nada.* [Dammit, calm down. This is nothing.]

MAMI: *Nos vamos a MORIR.* [We're going to DIE.]

PAPI (chuckling): *No nos vamos a morir, gorda. Es un poco de viento.* [We're not going to die, fat girl. It's just a little bit of wind.] (Gorda is a Cuban term of endearment that is sweet and has nothing to do with actual girth. Just like calling a female *"China,"* pronounced CHEE-nah, has nothing to do with China.)

MAMI (crying and burrowing inside her grocery bag–size Louis Vuitton purse for tissues and the St. Joseph medallion): Joo are an ASShole!

PAPI (indulging her, as usual): Why am I an asshole?

MAMI (beyond steamed): De fact dat joo even have to ask ees what makes joo one. Look at my face! My makeup ees gehtteengh all a mess because joo. Are. Here. Weeth. Dee. Turbulence! Dees ees from hell! We are all goheengh to DIE an' den joo have to ask why are joo an ASShole.

GIGI (noticing a fresh scuff mark on her brand-new Corkies, pigskin crisscross sandals with a cork wedge heel, purchased from the FBS catalogue; the French Boot Shop was a fabulous store in New Rochelle, N.Y., that always had all the latest styles): Mom, not to put too fine a point on it, but, uh, this was YOUR idea. Want a Tic Tac?

MAMI: No! I want to get the hell off de damn PLANE!

GIGI: God is protecting us! I wish he'd protect me from scuff marks. Dammit. These are brand-new sandals!

MAMI: I can't see anytheengh. What I can see ees dat we are here alone on a bumpy plane fool of fohkeengh Mexicans. Dees people do beesness weeth Feedehl Castro! Okay? An' das why I don' know how God feels about dees other people because dey are een cahoots weeth de dehveel heemself! An' we are outnumber-ed by DEM. Johs like our poor Jeweesh ancestors een de Holocaust. De women preesonehrs deedn't get der periods. Dey totally eh-stopp-ed, okay? Because of de cheer eh-stresses!

GIGI: Outnumbered. Stopped. Not outnumber-ed and stopp-ed. Gahd! HOW many years have you lived in this cute country now?

MAMI: Right, less focus on de American grammars an' een-veesahbl eh-skohf marks right now as we are crahsheengh to death in de horeebl Mexico desserts!

GIGI (looking in her compact mirror and attempting to reapply Yardley bubble gum lip gloss without getting it all over her face): DEH-zurt. Dih-ZURT is what you eat after dinner. And you love Mexico. You and Papi honeymooned here, remember? Acapulco, Los Cabos, Puerto Vallarta, Cancún . . . God, you know, my lips are so dry. I wonder if I brought my Kiehl's [lip balm #1]? There's nothing worse than trying to put lipstick directly on chapped lips. You have to have the moisturizing 'tude going on first, and *then* . . .

MAMI (talking to the medallion clutched in her Revlon claws): *!Ay, San José de Cupertino, ayuuudame!* [He-ee-lp me!]

PAPI: *Ana, por FAVOR.*

MAMI: *Coño,* Dahveed. Joo are a FEESEESHEEAHN! Make eet EH-STOHP!

Two weeks later in her private office on the mental ward, an unscathed Mami happily counted her cash profits from the Oaxacan black pottery. The psychiatrists and patients all really loved the stuff and coughed up huge big bucks for it. (Mami let some cash-strapped patients pay for theirs in packs of Kools.) The woman had survived the Mexican dessert storm and made out like a Cuban bandit. Works for her.

Was everything bad that happened to us Juban refugees really all Hitler and Kennedy and Castro's fault? The fact that my family and our friends had been scattered across the earth? Where was North Carolina? What was New Jersey? Where was my pretty ocean where Mami bathed me since I was five months old?

Miami and Miami Beach, I knew firsthand, were backwater honky pits. When my immediate and extended family arrived there in 1960 and 1961, these places were shitty little cracker towns with unbearable heat and humidity and no sea breezes and hideous pastel-painted houses ringed with shiny bushes where slime-green lizards lurked, ready to pounce and give you a heart attack. My parents and I lived in that charming setting for the first year of our exile. We three shacked up (sleeping together in one bedroom in a double bed, which I loved) with my *abuelos,* Zeide Boris and Baba Dora (they slept in a single bed in the broom closet); Tio Bernardo, my mother's fiery younger brother; Tia Ricky, his Raquel Welch look-alike Sephardic wife; and Joel, their son and my eighteen-month-old first cousin (they slept in a double bed in the second bedroom). The eight of us crammed into a tiny, cheaply furnished two-bedroom house that was part of a complex of ancient houses on Fourteenth Place in Miami Beach. The rent was $125 a month. We nicknamed the place Las Casitas Verdes, the Little Green Houses, for their yucky, chipped pea-green color.

My grandparents and Tia Ricky, who never worked, took care of me and Joel during the day. Papi was a bottle washer at the National Children's Cardiac Hospital, not far from Las Casitas. He also worked in their research department with rabbits and other animals. The people there treated Papi well and let him study for the foreign medical boards. Mami was a social worker in charge of medical eligibility at the Cuban Refugee Center. Tio Nano was a bank teller at the nearby Washington Federal Bank. They each earned about $65 a week, and gave all the money to Baba Dora, who paid the rent and bought the groceries (I use the term loosely, as we mostly ate eggs because they were nutritious and cheap). Tio Nano walked to work; the bank was only three blocks away. Papi and Mami needed a car, though, so they bought a $150 gray two-door 1950 Ford with a hole in the floor of the back left passenger side or the front right passenger side; nobody can agree on this. Wherever the hole was, I personally loved it. I thought it was *muy* fun that you could see the street running below you as you rode.

I whiled my days away by the Orthodox shul, a synagogue that was next to Las Casitas Verdes on the corner of Euclid Avenue. It was called Kneseth Israel Congregation. Twice a day, when the elderly American Jews made their way to and from services, Joelito and I sat on a cement bench out front, dressed in our tiny T-shirts, shorts, and sneakers, wailing in Yiddish, *"Goot yontif!"* Have a good holy day! Our family didn't join that shul, or any shul, for a while. First, my parents and I are Reform Jubanos, and second, we believed the Miami Beach crap was a temporary aberration and soon we would resume our normal lives in Cuba.

Two weeks into our exile, I turned three. We celebrated outside Las Casitas Verdes with a small day-old chocolate cake from Butterflake Bakery on Washington Avenue. Our recently arrived fellow Jubano refugees, relatives, and friends always hung out at our apartment, often sleeping over for days or weeks outside on the balcony (which was a pain when it rained; all the macho men would complain about their soaked *huevitos*—literally, little eggs, slang for testicles)—or else inside, sprawled on the couch or on the floor. The men stayed up till dawn playing *dominós* and smoking cigars, killing their anxieties, killing time, waiting for Fidel to fall, and he would fall, we all believed. I fell asleep to the men's sounds, the clacking *dominós*, inhaling the wafting cigar smoke and repeating the word *dominós* in whispers to myself over and over and over until I was dizzy and blacked out. Sometimes I varied it with all the *domin* words I could think of, some of which I had overheard our Catholic Cuban friends say: *dominós, Dominus, dominós, Dominus vobiscum, dominós, domino theory.* That last one I heard people say that President Eisenhower and later President Kennedy said. *Domino theory.* Clack-clack, click-click, puff-puff, more laughter, *Cuba si, Castro no, coño-coño-coño.*

Four months later, on Sunday, April 16, 1961, the day before the invasion of the Bay of Pigs, the radio began reporting that Havana was being bombarded by unidentified planes. The preemptive air strikes were said to be destroying Castro's military planes. Three U.S. airplanes piloted by CIA-trained Cuban exiles had bombed Cuban air bases and killed more than fifty people. (In reality, the few planes belonging to the Cuban air force had been dispersed, camouflaged, and replaced with obsolete planes left out to fool the exile fighters.) Dozens of Jubano families came to our apartment to listen to the news together. When the living room became SRO, the rest of the crowd gathered outside on the sidewalk, waiting, praying, smoking, sipping espresso, cheering, hoping. *Dominós, Dominus, dominós, Dominus vobiscum.* The next day the news reports were the same, and everybody was convinced Castro's days were numbered. Soon we could all go home! Mami would remind me of my beautiful air-conditioned bedroom in our Miramar house: all butter-yellow and white, with hand-painted furniture and polka-dot curtains . . .

The days passed, two or three, and bad news began streaming in. Castro's forces had easily overpowered our wildly outnumbered exile fighters in a bloody civil war that lasted seventy-two hours. It was a *desgracia,* a disgrace, and terrible failure. And it was all Kennedy's fault; that's what the Cuban exiles were saying. Kennedy had fucked us royally, our *mártires tan joven y guapos.* Our young and handsome martyrs. *¿Y pa' que?* And for what?

Back at Las Casitas Verdes, our crowded apartment, balcony, and front lawn gradually thinned out, getting quieter and quieter, until there were just the original eight of us left. Mami gave me a glass of sugary *café con leche* and a soft-boiled egg. I sprinkled salt on the egg and took a spoonful. Yellow and white mixed in my mouth. As I ate the colors of my bedroom, I kicked Joelito, who was hiding under the table and biting my legs.

"¡Me cágo en su madre!" I heard a man cry. I shit on his mother! *"¡Cabrón! ¡Que se vaya pa'l carájo!"* Bastard! May he go to hell! *"¡Nos jodió!"* He fucked us!

I disentangled myself from Joelito's teeth to see why the men were cursing and shouting in the living room. I mean cursing and shouting more than usual, since Cubans, unless they're expired, are cellularly incapable of lingual propriety, brevity, or reticence. They're the original anti–Strunk & Whiters. With words, like with shoes and money and lipstick and jewelry and parties, more is better.

I noticed my little toy lamb lying on the linoleum floor and picked it up, holding it to my chest. My grandfather, father, and uncle looked upset. The women were still off in the kitchen, murmuring softly and smoking their mentholated cigarettes, set apart from their men. I heard Tia Ricky warning Joelito, who was whining in there—what a baby!—to shut up or she'd give him *la chancleta,* the slipper. This is a typical Cuban mom threat, to beat you with her slipper/sandal/flip-flop/house shoe: *¡Te voy a dar la chancleta!* I'm gonna give you the slipper! Or the rhetorical *¿Tu quieres la chancleta?* You want the slipper?

The night came on, casting its long shadows across the room. It was very calm. My *tío*'s red hair, in a shaft of dying Florida sunlight, looked on fire. I watched and watched, fondling my lamb's frayed woolen ears. The triangle of men I loved most started weeping. I'd never seen grown men cry before.

"*Se acabó,*" they said.

It's over.

My parents' marital division of labor never changed much, even after Cuba. Papi was never directly involved in my life all that much. In some ways, he seemed beside the point. With the force of my mother's outsize persona overshadowing us all, he was relegated to a cameo role, which he expected, accepted, and excelled at. Papi kept the household humming; it was to him, not Mami, that the housekeepers gave their handwritten lists of necessary groceries and cleaning products. Whatever was needed from the drugstore—toilet paper, toothpaste, even Mami's feminine hygiene products—Papi took care of. If there were letters or packages to be mailed, Mami handed them to Papi, who also paid all the bills. If something in the apartment went wrong—be it a leaky faucet or a faulty lightbulb—Papi dealt with it or had to be the one to call a person to come in and fix it. Papi was both Mami's and my father, actually, though she was his pet.

My filial allegiance, however, remained steadfastly to my mother, and not one hundred sweet maids or nice dads would ever change that. The minute Mami walked in the door, the maid in question was abandoned like a worn-out toy. Mami the movie star was home! She glittered aloofly. The maid did not. Children love whatever's shiniest. Mami was glazzy, my word for one who is glamorous and jazzy. Indeed, falling under the bewitchingly romantic spell of an attractive but elusive one who enchants and beguiles you but who will never give you what you need would become an ongoing theme in my adult love life. My "expander" (never "shrink"—get it?), Dr. Adland, aka Gramps (because he was old enough to be my grandfather, though he claimed that treating me aged him prematurely into a geriatric), would say I was in search of a father figure, messing around with all these unavailable and fucked-up but fucking sexy older guys. I think I was looking for my mother with a penis and some passion, warmth, and intellectual and literary flair. I think I was following my father, too, the kind but remote model for all future lovers.

Meanwhile, I had a Cuban refugee child's work to do: learning *Inglés.* Someone had taken pity on my financially ruined family and blessed us with a small black-and-white Motorola TV. I religiously watched *Captain Kangaroo* and *The Lucy Show,* with Mr. Green Jeans and Mr. Mooney. I listened to Ella Fitzgerald records—my fave was "A Tisket, a Tasket." And there was an older girl who lived in our ugly brick building on Southeast Mississippi Avenue, Pamela, whom I met one day on the elevator. She marveled at my pearl earrings, remarking that she'd never seen such a tiny girl with pierced ears before. At the playground behind the dump that was our first real residence in the United States, all the other kids were equally mesmerized, taking turns touching my earlobes just to see if my pierced ears were real. I thought *they* were weird for *not* having pierced ears. Didn't all normal girls have them? These children are so childish and unsophisticated, I thought. They really need to get out more.

Papi and Mami always left and entered the building by way of the basement laundry room. They didn't want to socialize with the truckers, soldiers, and electricians—i.e., our neighbors—who assembled daily on aluminum folding chairs out front, drinking beer. It was hardly a matter of snobbery on my parents' part: What the hell did they have in common with rednecks, or them with us?

Plus, who knew, there might be another coup in Cuba, and soon we could put this fucking nightmare behind us and return to our normal lives. Meanwhile, until we bought our new "coats," Mami was still using Baba Dora's mink stole to ward off the cold, and Mami's diamond engagement ring, which she'd managed to smuggle out of Cuba, was the size of a Ping-Pong ball. It was all just too much to explain to outsiders. The *Americanos*—especially these charming neighbors of ours—would never understand it.

Some years ago, Mami, a civil servant, finally caved to the constant comment of people about her ring and placed it in a safety deposit box in the bank. Her replacement diamond ring is actually bigger than the original and so sparkly it could blind you. She keeps it that way with Windex. That is because it is a fabu-fake from Bijoux Terner, the *only* costume jewelry boutique that any self-respecting Miami Cubanita-Cubanasa would *ever* set a spike heel in. The Terners are old friends of my mom's, and they, too, arrived in this country with *nada*. Theirs is a really good story, how they saw gold in them thar fakes and built up this ersatz jewelry empire. The boutiques—there are several scattered across Miami—are the size of warehouses. You go crazy in there because there's just so much great stuff—Chanel knockoffs, amazing hair ornaments, all the grooviest, trendiest accessories you see in fashion mags—and the prices are so cheap.

But in those early days in Southeast, in D.C., what little we owned and wore was real. Poor = real; settled = fake. Very strange. So that when my new friend Pamela complimented my pearl earrings or my 18-karat yellow-and-rose gold ID bracelets (which all Cuban babies wear) as we dangled from the jungle gym outside or played inside with her many dolls—so many dolls! I was in heaven!—it was real. Pamela was the daughter of a sergeant stationed at Bolling Air Force Base in nearby Alexandria, Virginia. She was nice and pretty and blond. She gave me a bunch of books

I still own: *Pat the Bunny, Andersen's Fairy Tales, Tales from Grimm, The Cat in the Hat, Curious George, Ellen's Lion, Just So Stories,* and *Charlotte's Web.*

Early reading came a little easier to me because Mami had taught me to read in Spanish in Cuba using *La Edad de Oro (The Golden Age),* by our poet laureate José Martí, one of the only Cubans all Cubans can embrace. Martí was great. The cover of the book has a color illustration in the Renoir style, with a touch of rococo: a kindly and handsome man surrounded by beautiful, placid, elegantly dressed turn-of-the-century children surrounded by their fancy little imported dogs. Very French, very sugary. Mami repeatedly read me the long poem called "Los Zapatitos de Rosa" ("The Little Rose Shoes") about a rich mother and her daughter, Pilar. They go to the beach in a luxurious coach to see and be seen. Pilar is all decked out in her plumed hat and silken rose-colored slippers. As Martí describes the individuals Pilar and her mother encounter at the beach, he's in effect describing all of Cuban society and the distinct partition in class that existed during that period. There are *las señoras,* the Cuban, English, and French ladies who sit conversing *como flores,* like flowers, under their parasols. There's a creepy little rich girl named Magdalena, dressed all in ribbons and bows, who enjoys burying her armless dolls in the sand.

Pilar wanders off to the other side of the beach, where, Martí tells us, the sea is brinier, saltier. That's where the poor and the old people sit. Pilar meets a sickly, barefoot girl and gives her the precious *zapatitos de rosa* and the plumed hat. At first Pilar's mother is furious, but later she realizes what a good girl she has, and ends up also giving away to the poor girl her cape, her ring, a carnation, and a kiss.

Sounds mawkish by contemporary American standards. But as a Jubana child I just loved the poem because it made me cry and it was a reminder to share what you have with people who don't have anything. In Martí's world, that sharing is moral and voluntary. But in Fidel's world, that "sharing" is achieved through fiat, like a perverted Robin Hood having his way with those he deems *gusanos*. Worms. Worms like my mother and father. Fidel called everybody in my family that. I mean, not personally to our faces, but collectively to all who fled (and still flee) his regime. Was I too young to be a worm at age three? I may have been born with little rose-colored slippers on my soft little pounded veal cutlet feet, but now, exiled, the cutlets were bare. And I, a worm or the child of worms—which would have made me a worm or maybe a wormlette—had to learn to depend on the kindness of American strangers.

Pamela helped me make the transition into my second language by reading English-language stories to me and making me read others back to her, correcting me whenever I made a mistake. The most useful and helpful book she gave me was *The Cat in the Hat Beginner Book Dictionary;* the basic words—*camp, friend, happy*—were each illustrated and used in sentences. I loved that book. Or rather, I relied on it to expand my new lexicon. To master English was the most important and intense thing for me then. Learning it, getting it, using it confidently was as exciting and unfolding an experience as I could imagine, like Helen Keller making the connection between water and the word *water.* I deeply identified with Helen Keller, as I did with Carson McCullers's Frankie, and those two characters would become more significant in my life as time went on.

Emily Dickinson, with whom I share a birthday, wrote, "There is no Frigate like a Book/To take us Lands away," and truly, little made me happier than getting into bed with a pile of books. Whenever Mami went someplace she'd ask, "What do joo want me to breengh joo back?" and I'd always answer, "A new book." Unknown to any of us, at least consciously, I was already moving in a certain direction; my vocation was calling me by my name. When your child would rather stay inside reading books than go out and play kick-ball with the neighborhood kids—it's a sign. Mami would say, "Joo have to get out DER!" and I'd say, "Out *where?*" After all, I'd just been all the way up past the firmament with a dead girl named Karen who suffered, *really* suffered, even worse than we did for Castro, all for having red shoes:

"The bright sunbeams streamed warmly through the windows upon Karen's pew. Her heart was so full of sunshine, of peace and happiness, that it broke; her soul flew upon a sunbeam to her Father in heaven, where not a word was breathed of the red shoes."

It's hard to go from that height down to kids with unpierced earlobes fighting over who's safe and who's out on the street. Who cared?

Unlike me, my parents would always have problems moving between Spanish and English. As recently as two years ago, when a raccoon found its way into their suburban house, my broom-wielding father chased it through the living room yelling *"¡Vamos! ¡Vamos!"* while my mother shouted, "Speak to eet een Eengleesh!"

What happened to our old life? The balmy days of ease and Mami pushing me in a stroller on Saturday mornings with the tropical sun freckling our skin as we squint along the beach. Stopping at a café and kissing and hugging her girlfriends, Estela, Berta, the drop-dead beauties Anitica and Nedda, all with their babies, the sun sparkling off the women's bloodred fingernails and smiling red lips. Somewhere the handsome, strapping men were off playing clickety-clack *dominós,* puffing masculine clouds of earthy tobacco, punching the air with the pungent bouquet of cigar smoke. I sucked on the nipple of my guava nectar *con vitaminas* and drifted off to the song of surf, golden bracelets, and women's laughter; the perfume of Agua de Violetas, espresso, L'air du Temps, and imported cigarettes.

The American snow and ice were cold and bitter affronts in winter. "Ees like a constan' eensohlt," Mami said, shaking her head and looking heavenward from behind her big black Jackie Kennedy sunglasses as white snowflakes fell on the lenses. She furiously flipped the bird at the wan, gray sky.

"*Los Americanos* deal with it," I told her. "Their children make snowmen with carrots in the nose and scarves around the necks. Then they drink the chocolate *caliente.* With the mini marshmallows on top. Can we get those marshmallows?"

"Leesehn," Mami said. "Een de first place, dey do dos theenghs because dey are peasants! Dey theenk deyr Eskeemos! Dey don' know any better. Ees like de Wahndehr Brayt. Gross! An' dos marshmellons, dey are totally deesgohsteengh, so forget eet. Everytheengh here ees so WHITE! De brayt, de weather, de people, de melons. White, white, white, white! ¡Ay, Cuba!"

"Yo, bitch," I said. (I've always called my mother "bitch." Term of endearment.) "It wasn't MY idea." As in, to come here to this charming foreign albino country.

"Fohk Fidel! Okay? Say dat weeth me. Loud an' proud. Go ahead. Fohk Fidel!"

I removed one ever-present caramel-colored pacifier from my mouth. I say "one" because I wore an entire string necklace of rubber *tetes,* or pacifiers, each one in a different state of attrition. It was sort of like a security charm. One *tete* was designated solely for stroking my forehead, for example, and another was strictly for the right eyelid. We called the necklace *la gindaleja* (gheen-dah-LEH-hah), a phrase that has no real meaning and therefore can't be translated.

"I want that hot chocolate," I told Mami, twirling my pinky into the hole of the pinky *tete*. "And no more elephant ears. And I want peanut butter. Crunchy. Why don't you ever buy that?"

"*¡Coño!*" she cried. Dammit! "I told joo. Peanut butter ees made for de peegs! Joo know?"

"Well, I like it."

"No. Joo don'. Peanut butter ees what?"

"Peanuts."

"Right, peanuts. An' dey come from de farms! Like peekeengh de cottons. Ees from de Amereecahn Southern farms, joo know? Een de South! Like from de slaveries! An' den dey feed dat to de peegs! Das not for de leetl kiddies. We never had dat een Cuba. Never. Das johs for de peegs. De same ones dat eat de Wahndehr Brayt."

It took me decades to stop associating Skippy with slavery.

Whatever Mami loved and hated, it was to extremity, and I tried real hard to also love and hate the same. She was my golden girl on the moon, my redheaded Cuban Grace Kelly with pearls on her earlobes, around her throat, encircling her wrists and fingers, the milk-white fingers with the perfect long red fingernails, the glazzy gal who was sleek and smart and thrilling. Mami knew the answers to everything and was sophisticated, worldly, and effort-lessly beautiful, easily the most beautiful mother in the world, the very best woman. Oh, I was so in love with her. It never occurred to either one of us that it was possible to love her and still be sep-arate people. One pre-Gramps psychiatrist, an American named Raymond Band, whom I saw briefly in Washington, D.C., said I was conflicted about my own *individuation*. Meaning that he thought that I thought that if I broke away and became my own person—which is apparently what all the well-adjusted albinos in America did—something awful would happen. I'd feel as though I were abandoning my mother. And my father. Hadn't they been through enough bad things? I figured if I clung to them and never left them alone, they would be less sad about losing Cuba and less sad about life in general. We would be this tight little constellation of love, understanding, comfort, closeness, safety, familiarity, fluency, and support in an otherwise vast, indifferent, confusing, lonely, threatening, strange, and empty cosmos. To me we were like three wounded *mártires* who had endured something intense and unspeakable together in an unwinnable war in a far-off land that only we understood and could never explain to anybody who wasn't also a Cuban refugee. The American psychiatrist regarded me as though I were an alien.

I guess I was.

The D.C. summers were hot and wet, unabated by sea breezes. Mami didn't mind. She loved to bake—in the sun, not cakes. Being fair-skinned redheads, neither of us could tan. But we could freckle. The aspiration was to get so many freckles that eventually they'd merge into one gigantic tan. But from June to September my legs were covered in hives and assorted rashes from the chafe. Unlike Mami, I was "delicate." A sensitive Jubana hothouse orchid who reacted violently to all sorts of outer and inner weather.

"*¡Mira pa' esto!*" Mami said, examining my blistered inner thighs. Look at this! And then to my father, "*¡Has algo!*" Do something!

"*¿Qué coño tu quieres que yo haga? ¡Hay calor! ¡No hay dinero!*" What the hell do you want me to do? It's hot! There's no money!

"*¡La niña no puede vivir así! Mira pa' esto. A mi no m'importa un carajo. Compra un aire acondicionado, coño. O róbate uno, a mi que m'importa. Pero hazlo!*" The child can't live like this! Look at this. I don't give a fuck (what it takes). Buy an air conditioner, dammit. Or steal one, I don't care. Just do it!

"Okay."

My father went to Sears and bought an electric fan for $12. They put it in my bedroom. My hives and rashes cleared right up. Mami needed a target for her fury. Papi was the most convenient one to blame, as if he had caused our new impoverished circumstances with all its attendant problems and deprivations. Mami railed against Fidel Castro, she howled at the moon for the random unfairness of life, for the constan' eensohlt of snow, for the red rash on my thighs.

Later on we got a.c. and she kind of calmed down. It made her feel slightly less bereft, since in Cuba we always had a.c. in our houses. Just like we always had maids and silver stove-top espresso pots. Those are the culturally specific symbols of lifestyle continuity from one country to another. We didn't have much else here, though. We had one another, sort of. I hardly ever saw my parents, and when I did, we were rarely alone. They were either at work or resting in their bed after work or going out or throwing dinner parties for psychiatrists or resting in their bed from the dinner parties. I grew up surrounded less by children than by adult psychiatrists and other assorted health care professionals who were all Mami's new work friends. (Papi didn't really have friends except Mami. He was always too shy, withdrawn, awkward, and socially indifferent. I found those qualities alternately sad, infuriating, and touching.) I had always felt more comfortable and normal around much older people than I ever did or would around my peers—with a couple of caveats. The people I have always gotten along with best are straight women either my age or younger, married men of any age, straight men close to my parents' age, gay men around my same age, and blacks and immigrants of either gender and any age. After all, until I was sent in the fall of 1967 to a prep school in the fifth grade, my friends were almost exclusively black: nannies, children, and their families.

What I didn't grasp until Sidwell Frenzy, as I refer to that hellish eight-year period from fifth to twelfth grade, was just how racially segregated Washington, D.C., was. It was an unpleasant realization, one that I've really never gotten used to. Maybe that's the reason I identify so much with people of color, although I am technically white. I have a Caucasian Mexican-American friend, Mayra, who likes to point out that white *is* a color, one that "contains all of the visible rays of the spectrum." Well, Mayra went to Harvard. That's how those people talk. Acing the SAT verbals is just in their blood. But Mami says saying we're whitey crackers is wrong. Here's why: There used to be a discount chain of stores in the northeastern part of the country called the Cosmetic Center. One time Mami was shopping there and saw a fetching display of particularly colorful makeup. It was really hard to shoplift in there—visible cameras and mirrors everywhere—so she grudgingly walked up to the counter to "pay" for an olive eyeshadow and a rust-colored lipstick, two of her favorite "earth" colors.

"Oh no," said the cashier. "These aren't for you."

"Of course dey are, honey," Mami said.

"No, ma'am. See? This says it's for 'Women of Color.' "

"Right. So?"

"Well, it's for women of *color.*"

"Das right. I *am* a woman of colors. Peenk."

Unlike Mami, whose heavy-duty accent instantly gives her away as a ferner, as they say down South, I have no accent either in English or in Spanish, and besides, I look mainstream and un-ethnic. Therefore, I can "pass" as a whitey. Or someone peenk. Whatever. But in some ways, that only compounds how different I feel underneath the skin. I've had blacks, Latinos, and Asians treat me one way before they found out my background, and then act completely another way when I open my mouth and start rattling off in rapid-fire, perfect *Español.*

But integration is how Latinas and Jubanas have always survived. Paradoxes "R" Us. We're always integrating the European with the Indian with the black with the Russian with the Mongolian with the Spanish with the English with the Jewish with the Gentile with the Old World with the New World with the real diamonds with the fake diamonds. Making sense of what doesn't make sense is how we've survived historically. That and a lot of red lipstick and toenail polish. We, or at least I, feel invisible without it.

One time when I was monastically shuttered at home for a few days while on deadline for the *Washington Post,* I was under so much pressure to produce and scoop the *New York Times* that I forwent makeup! That tells you how bad it was. Anyway, once I'd filed my story, I had to come into the office to get edited. I took a long hot shower, washed and blow-dried my hair to frizz-free perfection (it takes six styling products to get it that way), and lovingly applied my makeup with an I've-missed-you! gusto. I caught sight of myself in the hallway mirror in my apartment as I was leaving and I literally said out loud, "Oh, *there* you are!" In other words, Jubanas are more "themselves" with makeup than without it. Something is terribly, terribly wrong if we skip it.

Which is why the first time I went to see Gramps without a trace of makeup (okay, I think I did have some Kiehl's clear lip balm on, but that doesn't really count), we got our multicultural wires crossed. We'd been together for a year or so and the trust was there, so I decided to do something really bold, really daring, really wacky: a naked face in public! But it was simply because I'd overslept. And you *know* how these Freudian fuckers will *never* pass up an opportunity to psychoanalyze your every last move. If you've overslept—and by age four or five, having been ringed by so many psychiatrists, I hardly needed yet another one to point this out—there's a *reason* for it and that reason usually has something to do with *resistance* or *avoidance.* I just didn't have the energy to go there. So I arrived right on time and, knowing I looked like a raging hag, immediately apologized to Gramps for my appearance.

"I'm so sorry!" I said. "This is the worst you'll ever see me."

"Did something happen?"

"Well . . ."

"I mean, what happened to make you feel so bad?"

"No!" I said. God, he *so* didn't get it. "I'm not wearing any makeup! You can't tell the difference?"

"What are you talking about?" he said. "I thought you said you'd had a crisis."

I had. But explaining it would involve an additional five years on the couch. Not that I minded—psychoanalysis was good for me, it was enlightening and permanently useful. But I quickly did the calculations. Let's see, $200 a session, at three times a week (Gramps would have preferred five because "It only gives you the weekend to hide"), at four weeks per month, at . . . forget it. No health insurance is that good. And while the 'rents were kindly picking up the slack on that tab, they had mixed feelings at best about my treatment and *no* mixed feelings whatsoever about Gramps's fee.

"*Coño, el tipo es un ganef,*" Papi remarked in Spandish, or Spanish and Yiddish. Dammit, the guy is a thief. Mami Dearest's aversion to paying for anything had long since rubbed off on Papi, who delayed "expansion" payments as long as humanly possible.

In return, I of course ran to my expander and reported exactly what my father had said. I thought that would be "helpful." Gramps, a hard-core German Jew, was, shall we say, not amused. The news unfortunately backfired on me:

GRAMPS: Goddammit. He called me that? Look. If there was a flood in your father's basement and I was a plumber he called to come fix it, we would expect that I would be paid for my work on the spot. Right?

GIGI: Right. Absolutely.

GRAMPS: If I completed the task correctly, I would stand there while he wrote a check. I wouldn't "bill" him for the future. That's how the world works.

GIGI: Right!

GRAMPS: Except in your parents' world, everything comes for free.

GIGI: Right!

GRAMPS: NO. That is WRONG.

GIGI: Oh. Yeah, you're right.

GRAMPS: Goddammit, I know I'm right. Stop agreeing with me.

GIGI: Okay. Sorry.

GRAMPS: I don't want you to be sorry. I want you to realize that this is YOUR therapy. I don't give a shit how you pay for it. What I do give a shit about is that I get paid in a timely way, and that is YOUR responsibility.

GIGI: Can't this one be an exception? I'm sure my dad will cough it up eventually.

GRAMPS: There are no exceptions. You either stand for something or you don't.

GIGI: Well, I mean, hello. I can't MAKE him write the fucking checks.

GRAMPS: You're going to leave here in a few minutes—the sooner, the better . . .

GIGI: Gee, thanks.

GRAMPS: You're welcome. And you're going home and you're going to sit down with your parents tonight and tell them the following: "Either Gramps gets paid NOW or we have to terminate treatment." You ask them: "Can we afford this? Because if

not, we have to make other arrangements for therapy for me."
And don't come back here until and unless you've had that
conversation and bring me that check. You got that?

GIGI: Holy shit.

GRAMPS: That's right, holy shit.

I was so freaked out about having such a blunt, non-
Jubanesque talk with my fantasy- and denial-ridden parents, for
whom explicit money discussions were more taboo than sex, that
as Papi finally, reluctantly, handed me the check, I felt the top of
my skull pop open and fly away.

In our next session, I described this bizarre, unfamiliar sensation to Gramps, who took the check and practically broke out
into a one-man "Hava Nagila" hora.

"Mazel tov!" he cried. "Your head's cracking open! Expansion!
It's about time."

But before your head can crack open, before anything, for that
matter, Jubanas are required by their families to learn how to
throw great dinner parties. As her Aunt Alicia told Gigi in the
eponymous movie I was partially named after, "Bad table manners, my dear Gigi, have broken up more households than infi-
delity."

Entertaining well is a skill that will always come in handy, like
being able to tweeze your eyebrows without a mirror or knowing
how to bikini wax at home without setting your kitchen and
nether regions on fire. Mami and the maid always served a Cuban
menu. Typically it was *arroz con pollo,* the Cuban version of the
Spanish paella, preceded by light appetizers, such as mixed
drinks and little bowls of cashews or plantain chips. (Cuban food
is a little heavy, and Americans experiencing it for the first time

always go hawg wild and overeat, so you want to keep the appe-tizer situation light.) *Arroz con pollo* is a well-seasoned casserole of saffron-scented and colored rice (usually short-grain Valencia rice, which is fat, creamy, soft, and smooth like risotto) with chicken, pimientos, olives, sweet peas, and whatever else you've got lying around. Mami served it buffet-style, with white or red wine, fried ripe plantains, guava-stuffed croissants, and a ro-maine salad with sliced raw mushrooms and sliced red onions and a creamy dressing. Dessert was always flan, aka *crème caramel,* embellished with some sort of berry purée on the side, and espresso.

Watching Mami in confident, recipe-free Cuban hostess mode is how I learned to entertain others. She'd show me how to set a table: The little fork always goes to the left of the big fork; the water glass is always to the left of the wineglass; multiple candles and cut flowers project beauty and power. Even though we were still pretty poor, making do with style has always been a quintes-sentially Cuban trait. It's a matter of dignity and good taste, of Old World manners and savoir faire. Mami always insisted on fresh flowers at the dinner table. Her favorites were and still are *margaritas,* daisies.

The first dinner party she had in Washington that I recall was in the late winter of 1963. It was set for 8 P.M. on a Saturday night, which is the traditional Cuban dinner hour and because one of Mami's favorite movies is 1933's *Dinner at Eight* ("Jean Harlow's hair was totally FAKE een dat plateenohm, but chee was so fohnny an' cute! I lohv-ed her!"). It was the maid's day off, but she and Mami had everything prepared well ahead of time, which is easy with Cuban food, since it's forgiving and tends to improve over time. At 8:05, the doorbell rang. Everything but Mami was ready; she was just getting into the shower. Papi had gone out to buy some wine or ice or something. In Cuba, if you invite people for eight o'clock, nobody with any civility or manners even *thinks* of arriving until nine-ish. Any sooner would be incredibly tacky, rude, and bizarre. Mami wondered why the doorbell was ringing and asked me to go see. I opened the door and saw a man and a woman.

"Whah, you muss be Gigi!" said the woman. She had a really weird accent and smelled perfumey. He was very tall and looked like Abraham Lincoln. "Aren't you juss precious!"

"Hi," I said, unsure of what to do next. They were all dressed up, but surely they weren't here for the party. It was only 8:05. So we just stood there for a couple of moments. Finally the man said, "Well, you think maybe we could come on in? Ahm Joe. This here's ma wahf, Josephine. Ahm a suhkahtris. Ah work at the hospital with your mama."

"Where are you FROM?" I asked.

"Mississippi!" they chimed. See, I *knew* they were ferners. They couldn't have been from Mississippi *Avenue;* that was the road we lived on.

"Are you here to eat our foods?" I asked.

"Isn't she adorable?" Josephine said to Joe.

"She's a pistol," he replied.

I of course had no idea what or whose vocabulary they were using—a "pistol"?—or what exactly to do. So I decided to let them in, but only because they weren't armed and dressed in guerrilla fatigues.

"Eat our nuts and chips!" I instructed them, and rushed off to Mami's room.

I found her in the bathroom, applying Maybelline liquid black eyeliner. The TV set was atop a rolling table that had been pulled over from the bedroom to the bathroom doorway so Mami could watch old movies while she was getting ready. A Kool cigarette burned in an ashtray on the countertop next to a glass of red wine.

"*¿Quién fué?*" she asked abstractedly, concentrating on the outer edge of her upper eyelid. Who was it? She was barefoot and wearing one of her fifteen bathrobes. I gingerly stepped around the snaking extension cord and TV set—Mami had built a fortress around herself—and sat on the toilet seat. The black-and-white movie playing had a sumptuous smiling lady who was dancing with a man who was singing about queens' tiaras and dressing in sables. I picked up a tube of lipstick and looked in the mirror while I applied some Cherries in the Snow.

"Two people," I answered. "What movie is that?"

"*Cover Girl.* Ees from de forties. I saw eet een Cuba. I like *Geelda* better, doh. Because dat was de first time dat anybody had ever seen a ray-hayt wehreengh rayt."

"But it's in black and white."

"Only leetehrally. Feegurahteevehly, joo can tell Rita Hayworth ees een a rayt dress. 'Put All de Blames on de Mames.' Das what chee sang. Dey joos to say dat rayt hair and rayt clothes don' match. Dat dey clash. But dat ees totally crap. Because Rita Hayworth, a Latina . . . See? De guy der, dat one ees Eugene Kelly. *Eugenio.*"

Mami always Spanishified English words and names.

"Eugenio Kelly is the man dancer and singer?"

"Jes. Was der somebody at de door?"

"Uh-huh."

"Who?"

"Two people. Nothing like Rita or Eugenio, though. These ones are from another country."

"Where?"

"Mississippi. Josephus and Joseph Steen."

Mami dropped her powder brush.

"WHAT? Joe and Josephine? Dey are here? Oh, my God! Eet johs can't be! Ees only eight o'clock . . ."

"Eight-o-seven, actually," I said, smacking my rouged lips together in the mirror. "Let me see some of that eyeliner."

"What? No! Go out der NOW an' tell dem dat I'll be out een a second. ¡Coño! Dees country ees reedeeculous! I hate eet!"

I slipped my bare feet into her emerald-green satin stilettos and shuffled off to the living room, still holding the eyeliner wand.

"My mommy isn't ready yet," I announced. "But you can put on my eyeliner on my eyes for a while." I handed it to Joseph Steen, shutting my eyes and extending my chin. "Anyway, don't worry. We have a lot of many foods."

I kept my eyes closed to allow the eyeliner to dry, and began singing what I thought were the words I'd heard to "Put Me to the Test" from Cover Girl: " 'I wear no farmer but to my charmer I hereby pledge my wall. In other words, I'm at your ceck and ball.' "

I stopped and opened my Cleopatrified eyes. My audience appeared puzzled.

I beat on, a little boat against the current: "I'm pretty sure my mommy and daddy will talk about you very much after you leave here."

I brushed my teeth, put on my yellow-and-white cotton gingham baby doll pajamas with the little red strawberries sewn around the neck and sleeves, and sat upon my small white bed. The Steens had given me a present. It was a book, A Child's Guide to Freud. It had an aggressively drawn series of black pen-and-ink sketches of a fiendish little boy and his family. I thought maybe it was about Freud Flintstone, written from the point of view of Pebbles or possibly even Bamm-Bamm. But instead, I found a story about a troubled child who gets taken to therapy by his parents:

"What Mommy and Daddy take you to is called a PSYCHI-ATRIC SESSION. The man in the chair . . . will call you a NEU-

ROTIC . . . Arguing back is called EXPRESSING ANGER. The reason the psychiatrist does not mind this is called THE FEE . . . Now you've gained INSIGHT. Call yourself AWARE . . . [Your parents are] SICK. You alone are FREE. You ALONE."

The last page had a drawing of the little boy with wings, flying above the earth, away from the sun. I wrote my name on the book in black Magic Marker and turned out the light. I pulled the blankets up to my chin and stroked my hair with my hair *tete,* wondering who writers were and how they wrote their books. I heard my mother's laughter from the living room. When we'd sat down to eat and cut into our *pollo,* blood had oozed out. Mami had never cooked before and she thought cooking *pollo* meant you basically warmed it up. Mrs. Steen kindly took over and made the blood stop by baking the breasts for another hour in a much hotter oven. Now I inhaled commingled perfumes, espresso, and cigarette smoke. There were no intellectuals or creative people in my family. Sure, they were well-trained professionals, but they were not *educated,* if you can see the difference. There was no one in my immediate circle who would encourage me to pursue writing, a fantasy profession I was envisioning. Writing was not considered a serious or even a heterosexual girl–appropriate endeavor. Even at this tender age, I was already well into my Latina programming: Don't achieve. You can be successful and intelligent— that way, your parents can bore the shit out of the neighbors at cocktail parties with all your many accomplishments—but you must always remember your place; if you're too competent and self-sufficient nobody will marry you or, God forbid, impregnate you afterward. This is really, really bad because marriage and motherhood are your raisons d'être. If nothing else, the Hispanic culture is traditional and hierarchical; the implicit rules stipulate that if you're a girl you can be smart *or* you can be beautiful. But you can never be both, so you better choose. I loved-loved-loved

Cherries in the Snow and eyeliner. I loved-loved-loved my books. In other words, I was emotionally my age but intellectually I was far older. This was a big problem for a four-year-old who already knew a lot about herself and what turned her on. Right now, that was okay. It would not be okay later, but later wasn't right now. So I kept my little writing fantasy a secret. Like masturbating, it was not something you did in public, and it was something that the more you did, the more you wanted to do.

Anyway, major (or even minor, for that matter) writing was really neither here nor there at age four! Soon I would be spending the summers with Mami on the air-conditioned ward in the mental hospital. At least they kept it cool in there. Jubanos may hate snow and ice, but we'd die without air-conditioning and cylindrical icebergs of compacted crushed cubes in our drinks. It makes us feel more sane and secure, reassured that everything's going to be all right, that we're in civilization. That we're *Kool*.

At present, surveying our diminished surroundings and the fact that we and the housekeeper were the only Hispanics I knew, I, Anastasia Romanov, I ALONE—with my recurring bad dreams of expropriation and death by water, my pretty pink angora party dress and my strange new Freud book, my treasured *gindaleja* of holey rubber *tetes*—preferred exile in Siberia or even Las Casitas Verdes with day-old chocolate birthday cakes, gross-out green lizards, and "simple retailer" Jubanos to this foreign English-dominant wasteland.

Good peanut butter, though. Crunchy.

CHAPTER THREE

Girl Under de Bed

✺

*S*o why the hell was I, barely four years old, going every day from our squalid walk-up tenement in Southeast Washington, D.C., to a mental ward? God knows I asked. Mami explained that unlike Americans (bad), Cuban parents (good) don't desert their children in summertime by sending them off to camp (death).

"Cahm. Johs de word ees really really bad, joo know? What ees cahm?"

"It has a P at the end, Mami. You say it with a P. Cam-*p*. You have to say the 'puh.'"

"Right. What ees cahm?"

"You get on a big yellow *guagua* [bus] and they take you to a place where it's got grass and a lake and little bunk bed houses and there's crafts called papier-mâché . . ."

"Hold eet right der. Papier-mâché notheengh. WHAT kind of houses?"

"With bunk beds."

"Right. An' what kind of place has houses like dat?"

72

"Is this, like, a riddle?"

"*¡Coño, qué bruta!*" Dammit, what a dunce!

"Are there any pens in here so I can draw?"

It's very boring for a young Jubanique child to be locked in a mental institution for eight hours a day, five days a week, even a mental institution that has housed such famous residents as Ezra Pound and, many notorious years later, John Hinckley. I felt as employed there as Mami was, except *one* of us was, how shall we say, involuntarily volunteering her services and the pleasure of her company to the medicated masses. It seemed as though all three hundred sprawling acres of the federal hospital grounds were just-kill-me-now beige and gray and brown. Mami's building, the Dorothea Dix Pavilion, smelled of Pine-Sol. It was all dim hallways and black and gray metal desks and cabinets in there, and no carpets so the linoleum floor was bare and cold. The in-patients, the majority of whom were heavily sedated, mostly sat around in loose clothes or bathrobes and defeated slippers on orange and turquoise vinyl furniture and watched TV in the "day room." Some tried over and over to relight long ago extinguished cigarette butts piled in smelly stale peaks in huge round purple aluminum ashtrays. The patients constantly hit up the staff for fresh cigarettes. At first Mami shared hers to be nice and to bond, but after a while she got hip to their tricks and began announcing, "Sorry, Meestehr/Mees [whatever their last name was], I only have enough for myself."

There were several other female social workers in Mami's wing, really nice American women who made a big fuss over me between seeing patients (and thank God they did, because I'd have died of psychiatric cabin fever otherwise). These were Mami's first real women friends in this country, and many of those friendships have endured. I'd bounce from one to another of their offices, providing comic relief and evaluating each lady's fitness as a potential surrogate mother to me. "Gigi's so *entertaining!*" they'd say. Hello, what choice did I have? *I'm four years old and stuck in a locked mental institution, for God's sake.* Throwing a tantrum will only . . . *get me locked in a mental institution!* It just doesn't get much more diminished than this for a Russian-Lithuanian-Polish-Cuban-Jewish *princesa.* What was I supposed to do? The revolution had put me—Gigi La Yiya Yiyi Yiyita Rebeca Beatriz Anastasia Lula Mae Luli Gorda China Muma Mumita Mamita Bruta Bobita Benes Andursky Anders—here. And my parents, too. Fidel Castro had put us all here, that *cabrón comemierda H.P.* [shit-eating bastard *hijo de puta,* son of a whore]. And my Juban culture—which stipulates that children never leave their parents, never never never never EVER, not even after they're married with kids (especially after they're married with kids)—had put me here. Shit! Unlike the patients, I seriously had better things to do. I could be bathing with my mother in Varadero beach right now, I could be picking tropical flowers in a beautiful park with my *abuelos* or eating my daily lunch—*bistec a la palomilla* (pounded, fried steak), *arroz blanco* (white rice), *frijoles negros* (black beans), and *platanitos maduros* (fried ripe plantains)—which could be followed by a delightful nap chaser in a hand-painted yellow-and-white imported crib and dreams of heiress-hood.

But nooo.

So. I could either have hourly heart attacks, like my Miami Cuban exile brethren have had for more than 406,888 hours of the last four-plus Castro decades, or I could make the best of things. That made me be resourceful, adaptable, imaginative, and self-reliant, whether I felt like it or not. What do they say? *Children are resilient.* Yeah. Right. If you're the tiniest one with the least power, you have to obey the bigger, stronger ones feeding, clothing, and housing you, even if they're completely flipped out. My parents were way too engrossed in daily survival and the effects of compounded transatlantic losses to be a Juban version of Norman Rockwell parents. (Actually, Rockwell, with his idealized, un-Hispanic, syrupy visions of Americana, is Papi's favorite painter. When he got into a shared private practice in 1963, Papi hung schmaltzy Rockwell prints of doctors and patients along the office hallways.)

Whenever my parents did pay attention and those magical, ephemeral, singular moments presented themselves, I grabbed on tight and savored those rations. *Give me love.* At night Mami sat on the side of my bed and in the lamplight read or talked to me—briefly. She'd start to rise and I'd grab her freckled arm and pull her back down to me. *I need more from you. Are we together? Are you my loving mother? Are we a family? Or not? I wonder why you won't be closer and more giving.* I thought I had articulated those things by my actions, but it didn't seem to register in Mami, or in Papi, for that matter, and it never would. Nor would it with the four Epic-Epochal lovers of my future romantic life.

Even now, my mother sounds a thousand times brighter, happier, and more excited to hear from me on the telephone or in E-mails than in person. Why is this? I was learning the ropes, the ropes of people's love limitations, and settling for what little I could scramble to get. My deduction: *I may have to steal the love I crave. Where can I go to shoplift love? Maybe I'm just not interesting enough to linger over, so I'd better work on that. I must be burdensome and not worth putting time and effort into. My needs must be too big and unwieldy, for meeting them takes too long and takes too much out of others, who have more important things to do with their time. My feelings about the state of my needs don't count or matter. I will always be alone in the world. Castro said if you need a friend, get a dog. Maybe I'll ask Mami and Papi for a dog for my birthday. A puppy could help me.*

A weird fatalism was taking hold within me. It wasn't logical. Could it be a chemical or genetic hand-me-down from my great-grandparents, who died anonymously in the death camps, piled high in pits like the extinguished cigarette butts in the day room on the ward? Zeide Boris, my father (though he's always denied it), and Tío Bernardo—each has had a predisposition to depression. How well you tolerate frustration—some of us break sooner than others—is an element of that predisposition. The messages I was picking up day-to-day seemed to trigger that fatalism, and converge into a submissive, sometimes self-destructive, melancholy: inevitability and those recurring dreams of death by water and of being ripped off. My Juban family had secrets, hidden agendas, unfinished business. What were those things? It was all very subtle and therefore hard to identify and to resolve. We certainly never talked about it at home, but I knew I had it, whatever "it" was. A masochistic malaise, perhaps? I was much too young to articulate it to anyone at the time, but I knew I had a certain . . . syndrome. It was Jewish and it was Cuban. Though it appeared to be at odds with the cheerful, lively, sparkly Cuban spirit called *el echar pa' 'lante*—which literally means the throwing forward, or the ability to push ahead—all Cuban exiles are sad underneath the skin. I was like Papi in the Jewish syndrome sense, Papi who metaphorically hid under the bed his whole life: ill-at-ease with anyone but family and intimate friends, resigned, sweet, mistrustful of strangers and yet too trusting of others' real motives, a little sad, always picking up the tab and not letting others do for him, perceiving the world as a dangerous, inhospitable place.

I once asked Mami why most of the Jews seemed so passively to submit to the Nazis. She said, "Because dey couldn't believe dat what was happeneengh to dem was *really* happeneengh to dem."

Dr. Raymond Band, my pre-Gramps psychiatrist, the one

who said I was conflicted about my own *individuation* and who regarded me as though I were an alien, once called me "a driven leaf"; I had no moorings. "I lived on air that crossed me from sweet things," wrote Robert Frost in his poem "To Earthward." Imagine, a New England Yankee farmer-poet-philosopher-anti-Semite who read at Jackie Kennedy's husband's inauguration, speaking to me. "I craved strong sweets, but those/seemed strong when I was young: The petal of the rose/It was that stung."

Dr. Band couldn't see that generations of displacement—first my great-grandparents, from their shtetls to the camps; then my grandparents, from Europe and Russia to Cuba; then my whole family, from Cuba to the United States—take their toll, and how. What little morsels my parents could offer me, and I don't mean materially, weren't sufficient. I remained hungry. So I looked for my mother and my father in other people. "A driven leaf." I wish. "I crave the stain / Of tears, the aftermark / Of almost too much love / The sweet of bitter bark / And burning clove."

Oh, if only I *could* live on air, like a leaf . . .

There were bright spots. There was a trip to the National Zoo on a wonderfully moody autumn day, my favorite time of year: Mami's deep gray cropped angora sweater (worn with a sleek pair of black ski-type slacks, black leather gloves, black ballet flats, diamond stud earrings, and a silver charm bracelet) matched the color of the sky. Her hair and painted lips were remarkably red. Papi bought me a paper bag of unshelled peanuts, much to Mami's gastronomic dismay, and we ate some and fed the rest to the elephants, whose skin matched Mami's sweater and the Washington, D.C., sky. We took pictures. Mami looked like a model.

Another time Mami rented a stroller and rolled me through the ample, quiet corridors of the National Gallery of Art. It was cool and dim, and Mami showed me the paintings she favored; most were of ballerinas. *"Mira,"* she told me. Look. There was Auguste Renoir's *Dancer* and Edgar Degas's beautiful ballerinas. I loved to see those graceful figures, especially from the back, where green, yellow, blue, or pink satin sashes were tied in bows around the waists of their white tutus. We passed some Monet landscapes. Mami liked them, too. Her tastes, such as they were, were definitely eclectic. At the gift shop, she bought two small Monet prints and a Degas called *Dance Class.* Mami had the three framed and hung the Monets in the bathroom and the Degas in her bedroom. She said that one reminded her of her ballet school in Havana.

More than a decade later, when I was in college in Paris studying literature, poetry, and art history, I saw similar paintings at the Louvre. The Degas dancers resonated more than even Leonardo's Mona Lisa, probably because M.L. is encased in a bulletproof box and is always surrounded by hundreds of tourist gawkers. At five-two, I could barely get a glimpse of anything but her mysterious, long, eyebrow-less almond eyes. Besides, Degas's ballerinas had always been Mami's picks; how could anything else dare to compete?

Even when life was no longer about fighting off the threat of starvation or eviction, my parents usually left me to my own devices. Maybe I'd gotten too good at playing the independent, happy-go-lucky, assimilated child. Maybe my interests (books, magazines) and strengths (writing) were too isolating and foreign to them. Maybe I was just really good at entertaining myself. Maybe the 'rents, as I refer to them with my friends, didn't get me because I was so different from them. Maybe my presence made them miss Cecilia and Cuba too much. Maybe I had the wrong anatomy. Maybe Papi only had enough tunnel vision energy to love one girl. Maybe Mami was competing with me and would never let me win.

At any rate, I thought I could overcome and rise above any and all of those possibilities, and outwardly mask my blue inner feelings, by embracing the *Gypsy* attitude: "Let me entertain you, let me make you smile, let me do a few tricks, some old and then some new tricks, I'm very versatile."

In Mami's office I found a yellow legal pad and government-issue pens. Since I could never go to cahm, where they have cool crafts like papier-mâché, I settled for sketching flowers resembling the ones I used to pick for Mami in El Jardín Botánico and El Parque Central. Living where we did in D.C., there wasn't much nature. Behind our crummy apartment building, though, there was a shallow running creek. My Tía Elisa, Tío Julio's wife, used to take me back there to wade when she and Tío Julio came to visit us from Miami. Tía Elisa was a Big Mama who loved to eat EVERYTHING, and I could easily coax her into taking me to the Safeway supermarket where she'd buy me all the exotic delicacies that Mami wouldn't: Wahndehr Brayt, Welch's grape juice, Swee-Tarts, and Skippy crunchy peanut butter.

Later, we'd sit by the creek (after Tía Elisa cleared away the empty beer cans and broken glass pint bottles, the cigarette butts and used condoms), and I'd extend one leg and then the other so she could roll up the bottoms of my pants. I loved the slippery smoothness of the hard rocks on the soles of my feet under the dirty water. Mami regarded my soiled feet after these immersions and shook her head, grumbling, *"Este país es una mierda."* This country is a piece of shit.

Bernardo—the Puerto Rican immigrant gang leader of the Sharks in *West Side Story,* not my uncle—would concur with that assessment. Mami had taken me to see the musical, my first movie in this country, and in the song "America," Bernardo is disparaging of the United States: "Everywhere grime in America, organized crime in America, terrible time in America." But Nardo's spitfiery girlfriend and fellow Puerto Rican, Anita, considers her native land ugly, diseased, and backward: "Puerto Rico, my heart's devotion, let it sink back in the ocean."

"Dat Rita Moreno ees *so* not Cuban, by de way," Mami said as we left the theater. "Not even close. So I cannot blame her. Puerto Rico ees a peet compar-ed to Cuba. Actually, eets a peet compar-ed to anytheengh."

"Maria is so beautiful!" I said, reaching for Mami's hand. "She looked pretty in the white dress with the red sash, like a Degas ballerina. 'I feel pretty, oh so pretty . . .' "

"Natalie Wood, jes," Mami replied, digging in her purse with her free hand for the car keys. "But lemme tell joo, honey, chees about as Latina as Jackie Kennedy, okay? Dat accent, what was dat? Mexican?"

" 'I feel stunning and entrancing, feel like running and dancing for joy . . .' "

"Watch eet. Joor about to run an' dance for joy eento de car."

I was. My visual world was out of focus unless I got right up close to it. I assumed everybody saw this way.

" 'For I'm loved,' " I continued, " 'by a pretty wonderful goy!' "

"Speakeengh of goys," Mami said, "who ees Jackie Kennedy?"

"What? Oh. President Juan's chic wife."

"Right," Mami said, handing me two sticks of Juicy Fruit. *"Abre 'l chicle."* Open the gum. Just as Cubans generically call any chocolate bar a "peter" (pronounced PEH-tehr, probably from *Peter* Paul Mounds, introduced in 1921), they refer to any brand of chewing gum as "chicle," from Chiclets. Mami never smoked a cigarette without having something else in her mouth, be it a drink or a mint or chewing gum. People who smoke any other way are "hard-core ahdeects an' deesgohsteengh." I unwrapped the powdery beige rectangles and neatly rolled each one up tight, the way Mami preferred. It's the only way a lady chews gum. I handed Mami one roll and put the other in my mouth.

"Dat Jackie has class," Mami continued, reaching for the lighter for her Kool. "Not as much as us, but eet's right up der. Joo can't be *fina* [refined] eef joor fat. Jackie knows dat. Chee can't take a bad peecture. Chee always looks perfect. Plus, I hear chee smokes behind clos-ed doors. We could be like seestehrs, practeecally."

The gum's sugary sweetness slid down my throat. It was so intense that I started tearing and nearly choked. We were southeast-bound from Pennsylvania Avenue, heading toward the Potomac waterfront. When we passed the Arena Stage on Sixth Street, Mami said, "Das where I want to leev."

"You want to live in a theater? Like, on the stage?"

"No, *bobita,*" Mami said. No, silly girl. "I want to leev *near* eet. Because ees theater an' because ees near de water. Das a great combo. Drama an' a reevehr. I lohvee."

At the Varadero beach in Cuba, where Mami took me swimming from the time I was a baby, there was only powdery-soft white sand underfoot, like talc, and the salty sea water was translucent and pristine. A thousand incandescent *caracoles,* seashells, and iridescent *concha de perla,* mother-of-pearl shells, studded the strand. Inside them I could hear the waves' voices; they murmured, moaned, and rushed.

"Mi sirena," Mami said, dipping me into the supple waves and back up. My mermaid. *"¡Cómo te gusta el mar y la playa, mamita!"* How you love the sea and the beach, little mama!

"Méteme, Mami," I always replied. Put me in, Mommy. I wanted never to get out of the water. Inside my tiny body some-

thing vibrated with gorgeous animation. It was the life force. Since life began in the sea, I was in my element in *el mar.* Four decades later, the fiancé called me *la canaria,* canary bird; that's what Stanley Kowalski called Blanche DuBois in *A Streetcar Named Desire.* Stanley said, "Hey, canary bird! Toots! Get OUT of the BATHROOM!" Blanche stayed in the bath for hours, singing and soaking. She said that after a long bath she felt so good and cool, so rested and refreshed.

"A hot bath and a long, cold drink always give me a brand-new outlook on life!"

Moi aussi, Blanche. I've never made an important life decision before I've taken a long bath or shower. Of course, *all* my baths and showers are long. (I mean, they're normal for me but they're long for non-Jubanas.) Maybe it's that religious cleansing motif, or maybe it's the return to the womb. Maybe it's because I, a sea girl at heart, was born on an island. Whatever it is, I need it to come alive.

"Méteme, Mami."

"So?" I suddenly heard Mami's voice say. "What kind of place has dos bohnk bayt houses?"

"What? Oh. Camp."

"Das right. An' what kind of cahm ees dat?"

"¡Ay!" I cried, impatient with her game. I got up from the hospital's antiseptic cold floor to stretch. "I don't *know* what camp is that."

"A *concentration* cahm!"

"Mami, I need to *interact* with children my own age," I said, repeating a new word I'd overheard one of the staff psychiatrists use to a colleague. "I do. I see them on TV and these American children are friends with other *children.* Do you like the flowers I drew?"

"Look," she replied, barely glancing at my artistry. "We are poor refugees because Castro made us dat an' he stole every-theengh johs like Heetlehr. I never knew my gran' parents because Heetlehr took dem an' he keel-ed dem. Who ees Heetlehr?"

"Castro's late father," I said, batting her exhaled plumes of Kool cigarette smoke away from my face. As the smoke was clearing I noticed a tape dispenser on Mami's desk. I rolled off some strips of tape and pressed my remembered Cuban flowers on her wall.

"Das right, das right," Mami said. "An' how deed Heetlehr get de poor Jews like joor poor dead great-gran'parents over to de concentration cahms?"

"In trains."

"Exactly! Wheech ees so seemeelahr to a BUS. Wheech ees how dey sen' de leetl keedees to go to de *cahm!* Now please. Go back een de day room an' play weeth de patients, okay? Ees fun for dem to have a leetl keedee hangeengh aroun'. Eet breaks up de monotony of being so krehsee. An' let me tell joo, honey, dees people are krehsee as hell."

Skippy, slavery. Papier-mâché, six million dead.

Mami forced me into school a year early. A refugee child, scarcely four, I was nowhere near ready. I could read pretty well, but my English was Spanglish at best, still a bit wobbly, and I had acute separation anxiety. Mami felt school was the only option. To her, I was becoming a total pill, whining about having to accompany her day after day to the mental ward.

But I had my reasons. It wasn't just the boredom and the fact that I was at least thirty-five years younger than everyone else and the grindingly gray, crushed cigarettey ennui of it all. The breaking point came when I spent a period of time on the children's ward. It was in a separate building—St. Elizabeths being a vast complex, almost like a little city unto itself on a hill, surrounded by high wraparound stone walls and guards at all exits. Mami figured sending me there would be a good idea since (1) it would give *her* a break and (2) hadn't *I* been the one whining about needing to interact more with people my own age? After all, she reminded me, I'd never specified that I had to *interact* with *unhospitalized* children.

Mami dropped me off, and a big orderly—all the orderlies there looked like black bouncers in white uniforms—locked the door behind her. The staff introduced me to their charges. The only differences between these children and the adult patients were that the children weren't allowed to smoke, they were *slightly* more animated, and they were considerably shorter. Otherwise, they were all the same: krehsee patients. This realization came as a blow, as I had (wrongly) assumed that children, krehsee or otherwise, just had to be less nuts than their elders. They hadn't lived as long and therefore hadn't had enough time to get really nuts.

¿Verdad?

Uh, no.

From my four-year-old point of view, mental illness was mental illness was mental illness. I played cards with the kids who could, and we'd read and watch TV and eat lunch, and it was fun whenever we got to go to art therapy because that department had great supplies compared to Mami's black and blue ink pens (but sadly, no papier-mâché). We also got to go to drama therapy, which I loved. They were working on a production of *A Raisin in the Sun,* and the director, a great big lady named Alfie Brown, who was a good friend of Mami's, gathered me up in her warm lap and let me stay there as she bossed everybody around. It was great. She even made an assistant go get me a Coke and a Snickers. I'd never had a Snickers.

"Does this have the peanut butter in it?" I asked Alfie hopefully, as I unwrapped the candy bar and sniffed it.

"No, sugar," she said. "But it has nuts. Whole lotta nuts."

"Some people call the patients 'nuts.' My mommy calls patients Mr. or Miss or Mrs. It's the respect sign."

"Your mama is right on."

"Are these PEAnuts?" I asked, pulling off a sticky piece of the caramelly candy.

"Mm-hm."

"My mommy says peanuts are from the slaveries, down in the South? That's why I have to wait for my Tía Elisa to come visit and then she gets me the Skippy. Cubans don't eat it if it's the crunchy peanut butter because Cuba's an island and we have to import and we didn't get the peanuts like the pigs in the American farms."

"You and your mama are somethin' else," Alfie said, clearly enjoying but also clearly having no idea what I was talking about—not that she could have.

I thought this was just about as good as being in Varadero with Mami. In this darkened, air-conditioned theater, safe in Alfie's ample lap, with my Coke and my chocolate bar, I watched patients of all different ages, races, and sizes, people who could barely connect with the real world—totally click by becoming other people in a play. It was astonishing. Like magic.

"Maybe I could be in your play one day," I told Alfie.

"Sure you could, sugar, and I bet you'd be real good. But this play here, see, this is about black folks."

Huh? I wasn't expecting that one. It confused me. I never differentiated race before. My Cuban godmother Nisia was black and my new Southwest D.C. neighborhood was mostly black (we'd recently moved into a high-rise overlooking the Arena Stage—"drama an' a reevehr"—and the dome of the U.S. Capitol) and my director Alfie here was black and many of the staff and the patients were, too. What did Alfie mean, "This play is about *black* folks"? I had never once heard Mami describe anyone by their skin color, except as a Cuban term of affection, *negrita, negrito*. The only thing I'd ever heard her say about blacks was that she wished she could be one because black women have the best skin and therefore age the best, and because she wanted to get soul.

"So I can't be in your play?" I asked Alfie.

"There's no part in the script for a little white girl, sugar. But that's only in this play. There's lots of other plays out there."

"Out *where?*" She sounded just like Mami. Joo have to get out DER.

"Lots and lots," Alfie continued. "You'll see."

Satisfied, I wrapped my arms around her and kissed her big cheek, feeling a surge of happiness splash up through me like a fountain.

It was either that or the sugar.

The artist Carroll Sockwell was one of Mami's patients. His family had had him committed as a young man because he was homosexual, or so he claimed. He liked to embroider and embellish things, just like Cubans. I met him a few times, and he left a lasting impression. Years later, Carroll would remind me of Richard Cory, a personage the poet Edwin Arlington Robinson wrote a eulogy for: "He was a gentleman from sole to crown, / Clean favored, and imperially slim. / And he was always quietly arrayed, / And he was always human when he talked."

None of Carroll's clothes or bathrobes was ever wrinkled or disheveled like so many of the other patients'. His hair and fingernails were perfectly groomed. He adored Mami and she adored him back. (All gay black men adore Mami. Actually, all gay men adore Mami. It must be the diva self-recognition.) Carroll was the only patient Mami ever gladly gave away her Kool cigarettes to. I guess therapists can have pet patients. He made amazing oil paintings and charcoal and pastel and pencil drawings in art therapy. The representational ones of faces that looked so sad and still and spiritually broken (I always imagined they were self-portraits) were in angular midnight blacks and blues. The paintings without people in them could be rounder, looser; voluptuous brown and mauve-gray horses' butts—or so I saw—with a very, very little bit of icy white and a sort of pinky ivory hovering in the bare trees around them. Later on, Carroll's work got abstracted and looked fully liberated. It was wildly colorful and geometric, and sometimes included torn pieces of paper, scraps of tinfoil, and tissue. Those canvases burst with bright turquoise blue, orange, yellow, green, and red—happy colors. They reminded me of piñatas and confetti, so naturally I preferred them.

Whether it was the Kools, the rapport, or the meds, Carroll began giving Mami his paintings as gifts, so many and so many big ones that they almost overlapped on the living room and dining room walls. It was the first "real art" we'd owned since Castro and his guerrillas took away everything we'd owned and loved and cared about. Carroll's pictures meant a lot to Mami. I think they made her feel less bereft. Less hurt. Owning good objects can make you feel loved and secure because they're tangible. Carroll was a paranoid schizophrenic and an alcoholic who painted like an expressive angel on jazz. It was strange to identify with him, but I did, in a way. Carroll Sockwell was able to give us something back in his art therapy that we had lost in Cuba: self-expression. And with that comes hope. We'd almost lost that, too.

After Carroll was released from St. E's, his painting career took off. He showed locally at the Phillips Collection, Washington

Project for the Arts, and in a one-person exhibition at the Corcoran; and at the Whitney Museum in New York, among other places. He and Mami stayed in touch sporadically throughout the years, and he came over to the house a few times with whoever his boyfriend was at the moment for drinks and now to sell her his pictures for just a few bucks apiece, pictures that became worth hundreds of thousands of dollars. Carroll would consume an entire bottle of cognac while he visited Mami. This was okay-fine by her. There is a veritable liquor warehouse in my parents' house. Some of the bottles are so old that they're dusty. This is because my parents have been given spirits, wines, and liqueurs as gifts for the past thirty years of parties. Being Jewish hosts, they rarely touch the stuff. One of the good things about being Jewish is the low rate of alcoholism; it's a supposedly genetic trait. However, none of our many Gentile guests could care less about that. They just love coming over and imbibing from our infinite alcohol archives.

"Carroll doesn't like espresso?" I asked Mami. We were in the kitchen fixing a tray of snifters, napkins, and bowls of plantain chips and salted cashews, while in the living room Carroll and his friend elegantly tapped their cigarette ashes in elegant white-and-gold-leaf ashtrays Mami had filched from a five-star hotel located near the Zürich Opera House.

"Espresso?" Mami said. "Are joo keedeengh? Ees, like, fohk de espresso. Breengh on de boozes!"

"Isn't he an alcoholic, though? Aren't you, like, helping to poison him?"

It was hard to look down at the beautiful tray—hand-painted and acquired during one of my parents' long and numerous holidays in Spain or Portugal or Israel or Russia or France or Italy or Argentina or Switzerland or Belgium or Puerto Rico or Brazil or Chile or England or Kenya or Mexico—and think, "I'm serving poison."

"Carroll ees an alcoholeec," Mami said. "But eef he doesn't

dreenk eet in here he'll dreenk eet *out der,* so what's de deeeefer-
ence?"

When Mami didn't hear from Carroll for long stretches she
said, "Prohahbly hees back to dreenkeengh. A lot. He's very dis-
turb-ed. *Pobrecito."* Poor little thing.

Carroll jumped off the Rock Creek Park Bridge to his death in
the summer of 1992. He was only forty-nine. Like Carroll,
Richard Cory also committed suicide on a calm summer night. At
least Mami got to keep Carroll's paintings, and that made her
happy. I found many of them gloomy and depressing as hell. But
that was beside the point. For bereft Jubanos, more is more.
Amassing and stockpiling *objets* is *toujours* where it's at. As T. S.
Eliot said so beautifully in "The Waste Land," "These fragments I
have shored against my ruins . . ."

After I'd spent a month or so on the kids' ward, a twelve-year-old
patient named Kevin had developed a crush on me. I think the
technical term would be *fixation.* He seemed harmless enough,
following me around like a smitten puppy, sitting next to me at
lunch and generally regarding me with lovesick awe. It was five
o'clock, the usual time for the orderly to unlock the heavy door
and let me out, as Mami would be waiting for me out front in the
car. I walked over to the glass-walled staff office and didn't see
anyone in there but the secretary, who said everybody was in a
meeting and to go wait by the door and someone would be by
soon. Kevin and I walked over there and waited.

"Wish you weren't leavin'," he said.

"I know."

"You're so cute," he said, taking my hand in his.

"I know."

"Can I have a hug?" It was more of an advance warning than

an actual request. He embraced me tightly, squeezing me against him. He felt bony and hard. I gently pulled away.

"Your hair smells like fruit," Kevin said. "Or candy."

"It's not fruit *or* candy," I corrected him. "It's Agua de Violetas."

"I don't want you to leave."

"I don't blame you."

"I said, *I don't want you to leave.*"

"Yeah, I got you the first time," I replied, rolling my eyes and wondering where my orderly was. I checked my pink sundress to see if it needed smoothing, then my sandals to see if they needed rebuckling or anything. Nope.

"Let's sing a song until the orderly comes," I ventured. " 'Big girls don't cry, big girls don't cry . . .' You know that one? 'Bi-ig girls do-o-n't cry-y-y . . .' "

Kevin looked at me strangely. It was an expression I'd never seen on anybody before. He suddenly extracted an office-size scissor from his pants pocket and lifted the hem of my dress with its slanted tip, laying the closed silver blades flat on my inner left thigh. The steel was shiny against my skin in the dim hallway light.

"I told you," Kevin said, slowly opening the scissors, "to stay with me."

"Okay."

"You gonna do it?"

"Yeah. Let's go back into the day room, okay?"

"Okay, baby."

He closed the scissor and slid it back into his pocket. I let him hold my hand. As we passed the staff office—only the stupid secretary was still in there, filing her nails, cracking her gum, and reading a magazine—I bolted away and ran inside.

The rest happened very quickly: me crying and hiding under a desk, the secretary screaming, a bunch of staffers running out of a meeting room, two orderlies subduing Kevin, who was kicking and screaming on the floor, another orderly carrying me out in his arms and delivering me to my mother, who was sitting in her car, oblivious, blowing the horn and pissed that I was late because we would hit rush hour.

At least I got to stop having to go there anymore, TYJ (thank you, Jesus). On the day of my liberation, I tell you I was happier than a Fidel-fleeing Cuban rafter making it intact to the U.S. of A.

Really, my deliverance was poetry.

The scissor incident is why I began school before I was really ready. It was the only ohpshohng. So Mami devised a plan, and by the time we met with the Amidon elementary school principal about getting me into kindergarten a year ahead of my peers—nursery school or pre-K weren't on anybody's radar at that point—it was basically a done deal. *La gringa* never knew what hit her; she was a total sohkehr.

"See, dees keed ees totally precocious," Mami explained. "I feel eet ees not an overstatemen' to say chees jooneek."

"I'm sure you think she is," said *la gringa*. "But children this age, developmentally, need a bit more time before they can successfully be integrated into—"

Mami squeezed my hand, pressing the pointy tips of her lacquered bloodred fingernails into my tiny palm. That was my stage cue—not to mention it hurt like shit.

"¡Ay!" I cried, yanking away my hand and wiggling out of my chair to climb into *la gringa*'s bony lap. (So far, so good. Anticipation, execution, all flawless. Not to put too fine a point on it, but I was pathologically cute.) I wrapped my arms around her skinny neck like an adoring kitten, kissing both her cheeks.

"I'm so so sorry," I told her, curling on her lap and remembering to repeat the word *so* twice whenever I used it. That was from the script Mami had concocted and had made me memorize before this meeting. The stage direction at that point said to look "soulfully" into *la gringa*'s eyes. I tried, but she was so tall and long that all I could get was two huge nostrils.

"I am so so sorry," I continued. "But you are beautiful! And I would say your pearls are very so so. I could not repiss!"

Mami was horrified. *So so pearls? Repiss?* Fuck. It was *reSisT.* Fuck. I'd screwed this up. And I was doing so well! Oh God, it was back to the mental ward. Mami mouthed the words *"¡Me cago 'n su madre!"* I shit on your/her/his/its mother!

The unaware *gringa*, however, found me fabulous.

"Mrs. Anders, you've got *quite* a girl here. I'd say your assessment of her acumen and readiness for kindergarten is astute."

"Joo are too kin'," Mami said, utterly relieved, feigning humility and refusing to pronounce the last letter of *kind.*

She always did like to have things her way.

Mami had to drag me into kindergarten. Soon after the scissor incident, I'd taken to hiding under my bed whenever it was time to take me somewhere new, afraid I'd be stabbed to death with an office instrument. Mami called me "de girl under de bed." It wasn't an accolade. She'd locate a stray arm or foot and maddeningly yank it and the rest of me out. I'd be like one of those little show poodles having a bad hair day, who doesn't want to perform, no matter what, and do a tug of war with my mistress. Being bigger and stronger, she of course would always win. I'd curse my head off at her in the car, accusing her of *reckless childhood abuse and endangerment,* another term I'd learned on the kids' ward.

"Joo know what?" Mami would say, "Das johs too damn bad."

I'd effortlessly conquered the principal *gringa*—charming and flirting with these authority figure types was one thing. Cultural flitting came easily; Jubana butterflies flit because that is their nature, not be pinned down is what they *do*. But actually lingering in structured, academic environs, staying on for entire "semesters" at a time—that was a whole different story.

My kindergarten teacher looked like Aunt Bee on *The Andy Griffith Show*, and while she was on the somber side, she was nice enough. She took me under her flabby wing, seeing as I was the only Hispanic child and the only student in the entire school for whom English was a second language. Eventually I began getting the hang of it. Finger paints—loved. Oatmeal cookies and milk—so-so. Skipping to square dance tunes with little blond boys—all right. Reading—fabulous, as long as I could hold the book right up against my face.

That last item got me sent to the school nurse for an eye exam. There was a chart with a series of capital E's across it in rows. The E's were rotated in different directions. There was a backward E, an upside-down E, and so on. My job was to sit at a distance from the chart and tell the nurse which way each E was pointing with my hand. I cocked my head right. I cocked my head left. All I saw were what appeared to be fuzzy black chopsticks tossed around like pick-up sticks on a white tablecloth.

"What do you see, Gigi?" the nurse asked me, jotting something down on a chart.

"I see . . ." I said, rolling my metal chair past her up to the chart, "I . . . Ohhh! They're E's! But they're all messed up. Look, this one's, like, dead. On its back like a poor dead baby bird. Aw. And this one looks like an M. And this one looks like the Hebrew letter . . . oh, what's that letter? My Zeide Boris's yarmulke has a letter like it on top in silver thread. That's *hand stitched*, by the way. *Embroidered in Cuba.*"

The nurse stared at me and blinked.

"Really," she said, humoring me. "Okay, go on."

"SHIN!" I said, smacking my forehead. *¡Qué bruta soy!* [What a dunce I am!] SHIN. SHIN SHIN SHIN. I love SHIN, don't you? It's the next to the last letter in the Hebrew alphabet, right before TAV. TAV is definitely the last letter in there, like in the Land of Milk and Honey. Baba Dora told me. I remember now. They put Zeide's yarmulke inside-out in the pocket where his linen handkerchief went on his jacket. He put his handkerchief in his pants pocket. That's how they got it past the guerrillas who stole our houses. Otherwise, who knows? Fidel might have used my Zeide's yarmulke as a Handi Wipe or something to clean his filthy mouth and ugly, smelly, greasy black beard or that gross dirt under his—eiw!—Satan claw toenails, like in the Hieronymous monsters. My mami showed me a Bosch print from El Museo del Prado, that's where that painting is, in Spain, in three pieces. Trip-tych. Spaniards killed poor Jews in the Inquiry and then all the poor Indians in Cuba to conquer Cuba."

"Gosh," she said. She had no idea what in the world I was talking about but she was riveted, in a bizarre way. This would become a leitmotif. (I've since fine-tuned it; few can handle me full-strength on the first exposure, so I always reassure people that it's the hardest one. After that it gets much easier.)

"Mm-hm, and *los* Ciboneyes, Guanahatabeyes, and *los* Taínos—those were the Cuban Indians," I continued. "We drink Malta Hatuey—well, I don't. That's a *cerveza*. Beer. It's named for poor dead Hatuey, the Taíno Indian chief. The Spanish priest burned the chief at the stake just like poor dead Juana de Arco because Hatuey wouldn't accept Jesus. Agh! It's like something right out of the summer concentration camps, if you think about it."

I stared up at the nurse, biting my lip.

"You need to see a special doctor right away," she said. "I'll write you a note for your mother."

"My daddy's a doctor and my mommy's a psychiatric social worker," I told her. "And I see special doctors all the time on the mental ward. That's why I'm here, so I don't have to go to see them."

The nurse looked utterly bemused and slightly alarmed. Her face turned pink and shiny.

These *gringas* were *muy* peculiar.

"Nooo!" Mami shrieked, throwing the nurse's note at Papi's head. *"¡No puede ser!"* It can't be! *"La tipa está equivocada."* The gal is mistaken.

"¡Tranquilízate!" Papi told her. Calm down! *"Esto no es nada, mi corazón. Esto nomás que's un* 'eye test.' *Se lo dan a todos los chiquitos."* This is nothing, my heart (another Cuban term of endearment). This is just an eye test. They give it to all the kids.

"Ven acá, mamita," Mami told me. Come here, little mama. She bundled me up in her arms and, nose tip to nose tip, squinted hard into my eyes, searching for a visible sign of my visual malfunction. From up close, Mami's sunflower eyes looked watery. I dabbed the outer edge of her left eye with the tip of my right pinky to catch a falling tear. Tears mess up your makeup.

"Eef chees blind, God forbeed," she told Papi, "ees because JOOR side of de fahmeely has all de bad eyes!"

"Perdón," Papi said. *"Perdóname. ¡Pero Bernardo usa espejuelos! Él mismo se llama* Mr. Magoo. *Así que házme el favor de ni empezar por allá."* Excuse me. Forgive me. But Bernardo wears glasses! He himself calls himself Mr. Magoo. So do me the favor of not even starting down that road.

Mami dropped me on the rented sofa—all our furniture was

ugly, cheap, and rented in that apartment and in its predeces-
sor—and flew out of the living room in a tearful huff, slamming
the bathroom door behind her and unsettling glass bottles of perfume and other toiletries. I wondered if my Jean Naté After Bath
Splash and Agua de Violetas were among the casualties. I looked
up at Papi. He wasn't saying anything. He just stood there, helpless. So I stretched out my legs and tapped the soles of my tiny
archless PVCs on the glass coffee table.

Tappety-tap-tap.

Tap.

"Gigi needs glasses," the pediatric ophthalmologist announced
after a series of failed eye tests in his darkened office. Instead of
letters up there on the illuminated screen, there were animals I
had to identify. All I saw were little blobs in different colors. Dr.
Kostenbader was elderly, soft-spoken, courtly, and balding, with
white hair. "She's quite seriously nearsighted," he told Mami. "In
her right eye, especially. I'm going to write you a prescription. You
should get it filled right away."

Mami took in the news with a façade of reason and acceptance. "Thank joo SO much," she said, taking the prescription
and neatly folding it in half. She stuck it in her caramel faux alligator faux Kelly bag. (As refugees we may have been poorer than
dirt, but being *Juban* refugees, we make do with panache and
good taste.) "We'll go to de optomehtrees' right away."

"She'll be fine," he said, reassuringly patting her hand.

"Chee EES fine," Mami said, pulling her hand away. The
abrupt gesture made the charms on her bracelet chime. I looked
up from above the top edge of my fascinating new fashion magazine (I had to read everything right up against my face, same with
watching TV) and rocked on my pony. Instead of regular chairs,

Dr. Kostenbader examined his patients while they straddled rocking horses.

"Can I keep this *Seventeen* magazine?" I asked him. "We don't get magazines. We're poor. Castro made us into exiled refugees and his father Hitler killed my great-grandparents."

"Uh, certainly, gal, take the magazine," Dr. Kostenbader replied, totally bewildered but polite. "And here, have a lollipop. May she have a lollipop?"

"Chee may," Mami said tersely, standing up. In her faux alligator caramel stilettos she towered over both the doctor and me.

"Do you prefer cherry or grape or—"

"I like guava," I told him.

"Guava?" he said.

"But I also like cherry, though. Grape, too."

"Not grape," Mami said, shaking her head and making a face, her index finger upright and wagging left and right like a manic metronome.

The doctor handed me a red lollipop and stroked my hair. It had been cropped short when we arrived in Miami Beach, but now Mami and I were letting it grow out. It was down past my shoulders, thick and straight as a curtain. The model on the magazine cover had longer hair, but I knew that if I gave it time I could be just like her, sitting on a gigantic pumpkin and smiling in a black beret, ivory cashmere fisherman cable-knit sweater, Royal Stewart tartan kilt, black Danskin tights, and black leather riding boots. (I had gleaned these infinitely engrossing details by reading the "On Our *Seventeen* Cover Girl" blurb inside the magazine while the doctor made us wait.) There were red and yellow and copper leaves all around the smiling *modelo Americana* with the long legs and no eyeglasses and such pretty clothes. We never had leaves like those in Cuba. We had, like, palm fronds.

Mental note to self: *Get the Crayola 64 pack and draw Ameri-*

can autumn leaves for Mami instead of the hibiscus and jasmine. Check to see if Hecht's has black berets, ivory cashmere fisherman cable-knit sweaters, Royal Stewart tartan kilts, black Danskin tights, and black leather riding boots. If so, see will Mami please go steal said items so I can be a beautiful and exciting cover girl instead of a short Cuban refugee with bad eyes. Then I'll be a better person and I promise I won't hide under the bed and Mami will love me more and read and talk to me longer at night and I won't die alone.

I took the lollipop.

"Thank you," I said, kissing the doctor's soft, sunken cheek. This lollipop and the *Seventeen* were really making my day, plus I got to be out of school. Mami took me by the hand and yanked me away. I implicitly knew why: Kissing Dr. Kostenbader meant I was kissing the Enemy. The Enemy because it was he who had just correctly diagnosed and officially confirmed my imperfection, thereby banishing any possibility of maternal denial. The doctor was, in Mami's 20–20 eyes, relegating me to a life of requiring glasses in order to see, which relegated me to eternal damnation not in hell, but in a much worse place: unmarried hag-hood. The place where boys don't make passes. The place where, because they don't make passes, boys don't grow up to become grooms who crush cloth-wrapped wineglasses underneath their black patent leather dress shoes as they stand beside you in a chuppah to be your husband.

My impairment was catastrophic, in other words. And so were my frame ohpshohngs, all three of them: plastic baby-blue cat-eyes (hideous), sissy pink plastic cat-eyes (even worse), or plastic black-brown fake tortoiseshell cat-eyes (don't even get me started). I settled for hideous.

Mami and I stepped outside.

"Mami!" I cried. "Mami! Look! The trees! They're so clear!

Look, they have a thousand green leaves! EACH! I thought trees were green blobs, like big bowls of lime Jell-O, but they're EACH, with the leaves!"

Mami was holding back from crying, her shapely white hand covering her red mouth. *Ay Dios mio,* she was probably thinking, plasteec baby-blue cat-eyes weeth a tafetán color champán wed-deengh dress?!? I want to dieee.

" 'I feel like running and dancing for joy!' " I sang. The last time I'd felt this high was bathing in the salty sea at Varadero beach. Although I have to say, that *Seventeen* find was also a life-altering thrill and right up there with Varadero. "And now I won't run and dance into the car because I can *SEE* it! I can see! The car's far away and I can SEE it from way across the street!"

Mami found herself a therapist the very next day.

Years later, I wrote a story for the *Washington Post* about little kids who wear glasses. People often wonder where writers get their ideas. You just write about whatever bothers you or turns you on or makes you wonder about the most and see if anybody else can relate. Usually, everybody else can relate. In this instance, I wondered if other parents also went plummeting into clinical depressions because their kids needed glasses. Guess what? Not one. The tiny kids I interviewed were all proud of their glasses. They said wearing glasses made them feel special and look brainy. (Then again, the ohpshohngs in children's frames are huge today. I've seen some French imports I'd be thrilled to wear myself, they're so adorable and hip.) Ronald Reagan used to say, "Trust, but verify." That's a good axiom for journalists. It also showed me that in her reaction to my needing glasses, Mami was, um, jooneek.

Mami insists that her sole concern was about how I would ever dance ballet and play sports while wearing glasses. Well, that is a scream unto itself, *n'est-ce pas?* First, I had no ballet inclinations whatsoever nor the body for it, and second, SPORTS? Was she insane? Her therapist said it was lucky to live in a time in history when there was a way to correct my myopia. He was right, of course, but for Mami that was neither here nor there, and if she could've found a way to blame Hitler or Castro for my—her?—catastropheec eempairmehn', she would have. Mami's difficulty dealing with exile status and the whole eyeglasses issue (as well as many other issues, some more "catastrophic" than others, that I subsequently presented her with) reminds me of Robert Frost's poem "Reluctance," in which he wonders why people have so much trouble going with the flow of things, and bowing and accepting the end of a love or a season: "The heart is still aching to seek, / But the feet question 'Whither?' "

Well, Mami's not exactly a Robert Frost kinda gal. I think she'd side more with poor dead Diana Vreeland, who famously said, "Style is refusal." Now *that*'s poetry. At any rate, from the age of four onward, whenever I was photographed for family pix, Mami always said the same thing:

"Quítate los espejuelos."

Take off your glasses.

Fortunately, Mami emerged from her funk long enough to force me into ballet—and later, drama—on Saturdays as a vicarious way to overcompensate. At first I hid under my bed. Mami had bought me a black Danskin leotard, pale pink tights, and pale pink leather slippers. Well, at least they were all Danskin, like the *Seventeen* cover girl model's tights. So that was something. Maybe you had to work your way up to, like, an entire American fashion outfit.

"Jool be johs like dos girls!" Mami said, pointing her chin at the Degas print on her bedroom wall. "Jool lohvee!"

"I like *looking* at them," I said, surrendering to the tights she was rolling up my little legs. "I don't wanna *be* them. I wanna be the *Seventeen*—"

"Of course joo do," Mami interjected, now straightening the slippers' elastic straps across my feet. She smoothed my hair, which she'd sprinkled with Violetas and spiraled up into a seamless French twist. We walked to the full-length mirror on her bathroom door and assessed.

"Look!" Mami said. "So cute! Anna Pavlova. Johs like me een Cuba. All de best girls danc-ed de ballets. Russian, just like Zeide Boris. An' French, like dat fohkeengh anti-Semite Coco Chanel. A classeec. Tradeeshon. Das everytheengh!"

"If I don't like that ballet and something bad happens to me there," I warned her in the car on the way to class, "you better get me and bring me home or I'll call the police. I dial the zero and the operator gets me cops. My teacher told me."

"Honey," Mami replied, exhaling her Kool smoke, "joo have to get out DER."

We greeted the friendly American *mulata* teacher, and Mami handed her a check, which the teacher swiftly folded and tucked inside her shallow caramel-colored cleavage. I was introduced to the class with a crisp clap of the teacher's hands.

"*Attention!*" she said in perfect French, a language that to me sounded foreign but not too. "*Voilà Gigi. Dites-lui bonjour.*"

"*Bonjour, Gigi!*" they chimed.

I had to admit, my name was perfect for this.

"*Buenos días!*" I replied.

Silence.

The "studio" was a makeshift affair. It was really a big multipurpose room located off the lobby of a high-rise (not far from our apartment) that had been converted into a dance space. There were long windows along the long parallel sides of the rectangular room, with horizontal mirrors hung on the walls and a wooden rail to hold on to, called a barre. It wasn't full-length, so there were rows of fold-up chairs compensating on either end. There was a portable record player on the floor, with various LPs: Igor Stravinsky, Piotr Ilyich Tchaikovsky, and Sergei Prokofiev. Off to the side, a skinny bald guy in a black turtleneck and jeans took turns playing a piano with a fat older woman wearing a flowered dress. We commenced our ninety minutes with some warm-up stretches and exercises at the barre.

"Mesdemoiselles," the teacher explained, *"les exercices à la barre sont le fondement du ballet. Ils sont comme les gammes pour le pianiste."*

I knew exactly what she meant, French words being so close to Spanish ones. The exercises at the barre are at the heart of ballet. They're like scales to the pianist. I knew what scales were for two reasons: Mami always pretended she was practicing piano scales, and Mami said the things that make a Cubana *Cubana* were playing the piano; pierced ears with gold, diamond, or pearl earrings in 'em; Agua de Violetas in the hair; taking ballet; having long manicured fingernails; and having a zaftig, curvaceous derrière. Except for the piano playing—which Mami said was out of the question for me because we couldn't afford a piano—and the fingernails, which never seemed to grow without bending and then breaking, I fit the description impeccably.

In ballet class I first learned the five basic positions of the feet and how to do *pliés* (*demi-* and *grand-*), where you bend your knees with your legs turned out. This wasn't too hard or bad so far. I wasn't sui- or homi- (cidal)—yet. Maybe Mami was right after all. Ballet class definitely beat watching crazy patients relight extinguished cigarette butts or having them press opened scissors against your thigh. Then I learned to make the sign of the cross (en croix) by moving my leg *fluidement* to the front, to the side, to the back, and then to the same side again. I asked, but the teacher said there's no sign of the Star of David. I said why not. She just smiled and shrugged. There were ripples of derisive chortles. I looked around. The girl in front of me at the barre, an olive-skinned blond with kinky hair and no ass whatsoever, had her head bowed and was kind of shaking with giggles. Same thing with the other two *gringas* in front of her.

"Those *glasses*," snickered No Ass.

Did she mean *me?* Was she talking about *my*, albeit hideous, glasses? Then I realized: I was the only person in there wearing glasses, hideous or otherwise.

"Lisa!" the teacher admonished. *"Silence, s'il vous plaît!"*

"Old lady Coke bottles," No Ass added, as giggles rippled across the room.

"Lisa!" the teacher repeated. *"Silence!"*

Lisa. What a stupid name for a No Ass. *Lisa. Lisa la Bicha.* Lisa the Insect. I tried to contain myself, but Jubanas can't. Let's not forget my little José Martí airport tarmac episode over my ultimately STOLEN little red trike, not that I'm bitter or anything; these puny *Americanas* weren't even armed like that gross-out Cuban guerrilla thug. Sohkehrz. I could take 'em easy, just like when sheer boredom back in Miami Beach Land forced me to pinch my dorky cousin Joelito, even if it meant getting bitten in return. Until La Bicha started up, I thought that only boys were the dumb and vicious people little Jubanas had to stand up to.

"Hey!" I told La Bicha, lightly touching her shoulder. It was surprisingly fragile and bony, just like my public elementary school principal's. Were all these *gringas* so bony like that? "Castro took my Papi's hospital! And my Zeide Leon's Cuban American Textiles! All the Andurskys got the bad eyes. We cry because we're *refugees in exile,* you ignorant *bruta.* That's why I'm blind! Because we cry so much! Because the revolution was BEYOND OUR CONTROL!"

"Mademoiselle Gigi!" the teacher cried, walking over to me and La Bicha. *"Silence, s'il vous plaît!"*

"Comme vous voulez, señora."

My spontaneous polyglot made the teacher laugh. Mami had prepped her ahead of time about Our Trageec Émigré Seetuation, so my outburst didn't totally throw her. The rest of the class, however, was culturally blindsided, dumbfounded. I stared them down, silently daring them to make a single other remark, about, say, my "primitive barbaric African cannibal" pierced ears, which is what a few of the neighborhood girls called them. I may've been out of breath but I was on a roll. The teacher told the class to take five and came over to me. She embraced me as I gradually paced my breathing to match hers, slow. She *plié'd* real deep and whispered that I'd do fine and I'd be a real good ballerina in time. I could see down the front of her plunging V-neck leotard. There was the tip of Mami's folded yellow check. The teacher told me my hair smelled good and was it a special Cuban shampoo? I rolled my eyes and explained it was the Agua de Violetas Mami'd put in there. She had no idea what I was talking about and asked me to tell her what "Aga di Volettis" was.

"A-GUA DE VI-O-LE-TAS," I said, correcting her. Jesus, she spoke French, what was her *problema en Español?* "It's violet water. It's not a shampoo. It's a hair perfume by Agustín Reyes of Havana, Cuba. It's a classic, like Anna Pavlov and Russian Mongolian eyes like Zeide's and Atticus Finch and Jackie Kennedy and Caca Chanel."

"Well!" she replied, just clueless.

She noticed my feet, which were stuck, more or less, in second position. She studied them, then took in the rest of my *placement* and *ligne,* the alignment or line of my body. Her gaze returned to my feet. As usual, my ankles were practically grazing the floor. Apparently, real good ballerinas' ankles weren't supposed to do that. She repositioned my feet as though they had actual arches and pulled my shoulders back, erect.

The native albinos were growing restless, presumably because the teacher was paying too much attention to me.

"Gi-gi's a squee-gee!" a girl trilled.

I didn't know what it was but it had to be bad, based on her tone alone. Did she just call me a refu-Gigi?

"Alb-i-no Esk-i-mo!" I cried. That shut her up. Hahaha.

I may have been the shortest, blindest, curviest, most flat-footed ballerina in the history of ballet, but I'd shown up, dammit. I'd emerged from under the bed and gotten out DER. These *bailarinas americanas* were NOT gonna make me quit. *I shall overcome,* I thought. I'd seen groups of black American people on TV sing that song, but they said "we." Mami had talked to me about the "ceeveel rights movemen'" and equality for blacks, but I felt I needed some civil rights of my own, even if I was a white Hispanic Jewish child. Didn't *I* have the right to be in the world without being picked on, ridiculed, singled out, put down, and laughed at just because I was different, from a different country and culture? Blanche DuBois said that deliberate cruelty is unforgivable, the one thing she had never been guilty of.

Deep in my heart I do believe I shall overcome. I am not afraid. I am not alone. I shall overcome the whole wide world around— some day.

Audrey Hepburn's Lulamae/Holly could have her Tiffany's. I had my beloved *gindaleja* waiting for me at home. I would suck the mouth *tete* tonight and suck it *hard*. I'd suck myself right into a *dominós, Dominus, dominós, Dominus vobiscum, dominós, domino theory* stupor. That was my plan. Good.

You have to give yourself things to look forward to when you are facing evil.

"Dios mio," Mami wailed. "Here we go again. De Andursky curse. First de eyes and now de feet. *Coño.*" Dammit.

The ballet teacher had told Mami that my feet were flat as matzos—I guess the Star of David inquiry kind of tipped her off as to my Jewosity—and that that would likely present a problem for the physical demands of ballet, especially if I planned to advance to dancing *sur les pointes,* on my tiptoes. Though that was years off—children don't go *sur les pointes* until they're eleven or twelve, when the bones of the feet are fully developed—Mami was going to see to it that I became a Russian Degas Cubana ballerina, Polish Andursky arches be damned. We went to see a pediatric podiatrist who recommended I be fitted with special shoes containing "cookies," physical arches to teach my unleavened feet to rise to (presumably Gentile) arch-hood. At the special shoe store, it was like reliving the eyeglasses nightmare; one pair of shoes was uglier than the next. Not a single black leather riding boot in sight. Rats. There were brown lace-up ankle boots that reminded me of the ones Red Skelton wore when he was playing Freddie the Freeloader and Clem Kadiddlehopper on his weekly TV show. There were also some black-and-white saddle shoes, only slightly less heinous. Since Mami and I couldn't decide which we despised more, we bought both pairs.

"Don' worry, Luli," Mami said, holding back her tears. "We can take off de ugly plasteec glasses an' de ugly ortopehdeec choos BEFORE de ceremony begeens an' den eet won' clash weeth de *tafetán color champán* gown."

"No glasses?" I said. "What if I trip on my way to the chuppah?"

"Joo won'," Mami said. "Ees not as eef joo have to see to get marri-ed."

I alternated wearing the two different choos to school, depending on my outfit, praying to God no one would notice.

"Combat boots!" screamed a boy named Shawn the first day I wore the ankle boots. "Haaa. Hey y'all, Gigi's got combat boots! Com-BAT, com-BAT. You a soldier in 'Nam, you tropical termite!"

It was during recess, and I was swinging on the playground swings. When I heard the taunt I dropped my head and turned my feet inward, trying my best to hide them from view as I swung backward into what I hoped would be oblivion. I felt hot and sweaty, and the lenses of my hideous plastic baby-blue cat-eyes got steamy. I could feel blood rising up through my neck to my cheeks, flushing them. Oh, I was on fire, all right. It was bad enough to have been victimized by a fatigue-wearing bearded bully who called me and my family *gusanos*, worms. But now I was stranded in *el exílio*, in exile, and I was a TERMITE? This was way too much entomology for me. I mean, okay, insects are the most successful survivors and the dominant life form on Earth. And I may be a tropical little refugee girl with problems. But I'm not vermin, *coño*.

When the bell rang to go back inside, I watched for Shawn and followed him. As we were going down the hall to our classroom, I stopped and let him get a couple of yards ahead of me. Then I made a run for it, tackling the fucker. "I don't know about 'tropical,' " I yelled, "but how about a nice Hawaiian punch?" I smacked him hard upside the head and kicked him with the indestructibly ugly square *pointes* of my com-BAT termite-*gusano* boots. Maybe this would help my nonexistent arches rise; you know, by flexing them. Everybody gathered around, staring. Shawn knew better than to strike a girl, even in self-defense, so he mostly shielded his face and groin from my 'Nam—whatever that was—boots, screaming, "Stop, termite! Termite, stop! Go back to the tropical war where you came from!" He even tried to snatch off my glasses. I told him to go ahead and take them, I knew they were hideous, just like his face, and if I couldn't see his face I'd feel a lot better anyways.

The teacher, whom Mami had earlier briefed about my "podeeahtreec catastrophee," emerged and dragged Shawn away by the ear. He had to sit in the corner for thirty minutes. The teacher took me aside and explained that Shawn's older brother had been sent to fight in a war far away in a place called Viet-NAM. She said maybe that's why Shawn was so sensitive about it. I asked her what tropical was and she said it was where Cuba is. I asked her why he called me a termite and she said that there are a lot of termites in the tropics and also probably Shawn's house had bugs. I told her I was sorry about all that but if Shawn insulted me again he'd get a swift kick in the *huevos* and the *culo* with my ugly revolutionary combat boot. I bounced my index finger on my butt to demonstrate *culo* and I let her fathom the definition of *huevos* on her own.

The teacher made Shawn apologize to me in front of the entire class by repeating after her: "I'm sorry for what I said to you. It was wrong. I am ashamed of myself. You are not a soldier in the war. You are a nice, bright girl. There are no termites in this school. There are no termites in this school. There are no termites in this school."

Nanette, my friend who sat behind me, tugged my braided ponytail in feminine solidarity. She loved playing with my hair, now halfway down my back. Nanette liked to unbraid my ponytail and marvel at it coming undone and loose, then she'd rebraid it. "Black hair won't do that," she always said, meaning that if you undid her assorted braids, her hair would just sort of stand there, sticking out and defying gravity. Nanette wasn't the only black girl who loved playing with my hair; all my black and biracial girl-friends did. Actually, I had only four white friends, Annie, Holly, Rena, and Mara, and they never played with my hair.

There were these two or three older black girls who for three days in a row yanked my pink mother-of-pearl hair band off my head after school just to torment me while I was playing double Dutch. "Playing" may be stretching it; everybody knows that white girls, even hip-swinging Jubanitas like *moi,* do not have what it takes to be a double Dutch diva: an amazingly precise combination of fast feet, strong arms, steady rhythm, and sheer stamina. I wasn't coordinated or quick enough to be a rope turner, much less a jumper. These girls—Rhoda, Fairon, Zena—were SERIOUSLY gifted, no less so than the synchronized Busby Berkeley chorus girls in *Gold Diggers of 1933* or Esther Williams & Co. in *Million Dollar Mermaid,* two old movies Mami and I had watched on *The Late, Late Show.* So I got to recite the jumping rhymes—in iambs and in trochees, Shakespeare would have been proud—while the double Dutchers dazzled me with their progressively tricky flair and dexterity:

So, so, suck your toe, all the way to Mexico, once you get there let it go, so, so, suck your toe, spell it out so you can go now M, E, X, I . . .

Don't say ain't or your mother might faint and your father might fall in a bucket of paint, now I'll betcha a hundred dollars that you can't do this, now close your eyes and count to ten, one, two, three . . .

I got ants in my pants and they're making me dance now one, two, three . . .

I loved being included, if only as a sideline barker. *Mis negritas* and their families were as close as I would ever get to living in a Latino neighborhood, a place where you'd feel at one with the rest because you belong, where you are part of each other and your shared culture. As for those bratty older girls who messed with my hair band, on the third consecutive day I stopped right in the middle of chanting "now close your eyes" and marched straight to the principal's office to report the perps. They never messed with me or my lovely hair accoutrements again.

Otherwise, I've always loved having people's hands in my hair. Sometimes a good shampoo, like the wonderful aromatherapy ones I get at my Foxhall Square hairdresser's in D.C., or a head rub, like the amazing ones my Raleigh, North Carolina, masseur Darryl gave, can be better than sex, and last longer. Must everything really always be about genitalia, end of story? How limiting is that? Cubans are long, luxuriating lovers, the delight is in the lingering. We're not "efficient" in the North American microwave sense of in-timer-out-eat-over. Many's the time I've begged the un-Cuban fiancé to massage my abnormally large head, but he turns my request into a double entendre, touches my head for maybe thirty seconds, and we're right back to my hand on his penis. Sigh. If I were a lesbian I bet I'd get my head rubbed because my girlfriend would understand that pleasure and the sensual power of all the skin, and she'd do it for the sake of doing it, and not be in a hurry to get it over with to appease me and rush me along so we can move on to the Main Event of her penis, her penis, her penis. (Well, she obviously wouldn't *have* a penis, but you know what I mean.) I checked this out with a bona fide lesbian friend and she shook her head.

"Forget it," she said. "It's no better or easier over here."

Mami says that during the Cuban Missile Crisis, which she claims she knew would never come to a head, we discussed it at the dinner table every night while it lasted. I really don't remember. What I do recall is having a *very* busy TV viewing schedule in 1962. Between *Hazel*, *The Lucy Show*, *Red Skelton*, *The Flintstones*, *Ed Sullivan*, *Jackie Gleason*, *Lassie*, *My Three Sons*, and *Dennis the Menace*, I was alternately annoyed and transfixed watching news stories that kept interrupting my regular viewing. When John Glenn orbited the Earth in February, an event I found only mildly interesting, the Jewish joke around the house was, "Beeg deal. If joo have money joo can travel." I thought it must've taken brillions and gazillions of dollars to get that astronaut up there, 'cause my teacher told me we were in a space race with the Soviet Union, and Mami told me we hated the Soviets because "dey are een cahoots weeth de dehveel [Castro] heemself."

I wrote Jackie Kennedy's husband a letter, with Mami's proofreading help and Papi's four-cent stamp, asking why were we spending so many dollars on space when my Cuban family and all the other poor Cubans in exile were in greater need "because you, Mr. President John F. Kennedy, let us all down last year during the Bay of the Pigs. Really, that wasn't a good idea or nice. You should say you're sorry and will fix it. My Mami and Papi say now Fidel Castro, Adolfo Hitler's boy, thinks he's so big! And all the Cubanos here will not vote for you anymore. They'll go Republicano. Love, Gigi. P.S.: Mami says to tell you, 'You blew it, Juanito.' "

Mr. Jackie Kennedy wrote me back a few weeks later on the most beautiful heavy ivory linen stationery I'd ever seen. ("Das notheeng compar-ed to Papi's an' my weddeengh eenveetations," Mami said with a dismissive sniff.) The president said that he understood my position and the plight of the Cuban people and that he'd try to do better by us. But he said the space program would go forward regardless of the price because it was integral to the United States' interests in this and other hemispheres.

I didn't quite get that last part—"integral"? "hemispheres"?—
but I got the gist.

"We're not gonna get any money," I told Mami.

"Dat feegurs," she replied, exhaling her Kool smoke and criti-
cally examining her long painted fingernails. It was time for a
fresh manicure. "But ders always enough money for Jackie's hats
an' choos an' for Feedehl to steal from us. Das de way eet goes."

One day before Mami's August 6 birthday, Marilyn Monroe
died. I didn't know who she was but Mami explained that she was
a sad actress with fake bleach-ed hair, who didn't wear *calzoncillos,* underwear, and didn't bathe often enough and was Juanito F.
Kennedy's and everybody else's lover. We watched the report on
TV from Mami's bed while she preceded her "tehrahpy," or minis-
tering to her nails, by trimming mine. She cut straight across the
edge of a tiny nail almost to the opposite end and let me pull off
the rest.

"Joor nails," she grumbled, shaking her head. "Andursky
nails. Soft an' chort an' bendy. *Coño.* Benes nails are hard and
long an' strong. *Mujeres Cubanas* have to have nice nails. Ees our
trademark. I'm goheengh to buy joo gelateen. Knox."

"Fort Knox, I heard of that. It's where the gold is."

"Right but dees Knox ees like Jell-O," Mami said, blowing the
emery board dust off my nails. "Joo eat eet. Ees goo' for de nails."

The collective bouquet of acetone, nail polish, espresso, burning Kools, and dead movie stars without underwear had me
dazed, entranced. I felt like a little bug on a drug. But not a gross
bicho like a *gusano* or a, God forbid, termite or a *cucaracha* like
poor dead Franz Kafka. Maybe like a *mariquita,* a ladybug. They
don't hurt anybody and you should never kill one because they're
good luck. I could definitely use some of that.

CHAPTER FOUR

Huevos

A Jubana never forgets her first WASP. And Valerie Ogus, Vancouver-born second wife of D.C. insurance magnate Walter Ogus, was mine. This exotic creature was related to us by marriage; Walter was Baba Dora's first cousin. Like Baba's, Walter's origins were Lithuanian. His family migrated during the turn of the century and settled in Boston and then in Washington, D.C. By the time I met the Oguses, in the early or mid-sixties, they seemed über-American, almost Brahmin. Walter came from a generation of Jewish men whose habit was to marry Jewish women first, make a ton of money, get divorced, and marry trophy shiksas. In those latter unions the men take on more WASPy attributes than the women do Jewish ones. Maybe WASPy attributes are more fun to take on. At least there's no chronic hand-wringing, depressive, brooding, fun-killing, over-introspective Woody Allen neurosis to deal with, and certainly Manischewitz is never involved. (Although you will have to learn how to drink unsweetened beverages and handle power tools/gardening supplies/the heartbreaking fact of no dessert.)

But I suppose that if you grow up being embarrassed by or anti-Semitically harassed over your Old World *Fiddler on the Roof* shtetl Jewosity in this young Christian country, if you feel third-rate, then one way to diffuse it is by marrying out. Not that there's anything wrong with that! Hey, Gentile lust has worked huge for Ralph Lauren. And you don't even have to be a bazillionaire Polo pony designer Jew to have it work for you. Take my brother Eric. Born just three months prior to Juanito Kennedy's assassination, Eric and his generation are Doing It, too, and making it work for them, sort of. Though she officially converted to Judaism and the wedding was Jewish, my now ex-rated sister-in-law Roberta, a Minnesota farming gal, is a WASP, born and bred, through and through. And Eric is a true blue Jew, although he makes his living through a chain of fancy Southern California restaurants selling pork. *Trayf.* (At least I think that's the spelling. As my rabbi Bruce says, when it comes to transliteration of Hebrew to English, it's tough getting scholars to reach a consensus. It's a really great word, though. I love *forbidden*—unkosher—food. It's as key to the Reform Jubanique vocabulary as *puerco asado,* roasted pork loin. We see no conflict there.)

At any rate, the Oguses were warm and generous, two adjectives that frequently and tragically elude garden-variety WASPs who like gardening. The Oguses just seemed—and were—wonderful, fascinating, attractive, vivid, and really, really, really rich.

The main attraction of them for me was Valerie. I'd never seen anyone like her. Tall and slim and perpetually tanned from playing golf and swimming, she had cropped blond hair highlighted at the Lord & Taylor salon in Chevy Chase, Maryland. She had small Windex-blue eyes, a ruddy complexion, fabulous cheekbones, and wore bright pastel pink or coral lipstick. She drank like a *pargo,* a red snapper, smoked like a *chimenea,* and laughed like a

man, deep and hoarse. I think you'd call her "handsome." A former buyer in New York for Bergdorf Goodman, Henri Bendel, and
Bonwit Teller, Valerie had impeccable taste, especially in clothes.
No uptight, portly Talbots matron, she, nor remote, anorexic Truman Capote swan—the only two WASP subspecies I'd seen up till
then on TV and in magazines and catalogues. Valerie had inborn,
unteachable, eternally untrendy chic, the kind mere mortals can
aspire to but never quite grasp. To me she was Lauren Bacall and
Katharine Hepburn in one, relaxed and upbeat and no-nonsense
in her effortless glamourosity and decisive femininity, but with an
eclectic, interesting, witty, eccentric edge that precluded her from
ever being an Ordinary Gentile. I lived surrounded by Gentiles in
my Southwest D.C. neighborhood, of course, but not like Valerie.
I seriously doubt Valerie could relate to my double Dutch Baptist
divas, and vice versa.

Valerie always reminded me of the kind of gal you'd have seen
dancing and drinking it up and laughing the night away (with her
head thrown back) at Truman Capote's Black and White Ball.
Only if Valerie had been invited, she probably would have skipped
it. Society types, poseurs, and society-type poseurs weren't her
bag. She preferred giving sumptuous yet casual and very intimate
dinner parties at her palatial apartment by the Shoreham Hotel,
where the drinks were strong and the entrées never varied.
Valerie's signature dish (she did all the cooking) was a big fat juicy
hamburger. Sirloin only, chopped special by her Chevy Chase
butcher, served bunless, like steak. French fries. Caesar salad,
tossed by a designated guest. All served on individual trays and
consumed in the living, never dining, room. Frank Sinatra
crooned in the background about city girls who lived up the stair,
with all that perfumed hair that came undone . . . And about
blue-blooded girls of independent means who'd ride in limou-
sines, their chauffeurs would drive . . .

"I was always plaster-ed—an' how—whenever I was der," Mami recalls. "Eet was a meestehry to me as to why."

Eventually Papi figured it out. He explained it was from mixing highballs with *vino*. Mami wanted to do what Zeide Boris did, which was to have *un wiski*, or Scotch in Cubanese. So Mami would start with cocktails, usually Scotch on the rocks. Then she'd drink red wine with dinner. By the time coffee was being served, Mami was smash-ed. In keeping with certain WASP predilections, Valerie served dessert—usually high-quality vanilla bean ice cream lightly topped with an insouciant toss of ground espresso beans or grated bittersweet chocolate—only when my parents, with their Cuban sweet teeth, were present. Heavily sugared joo-name-eet is to Cubans as hard liquor is to WASPs. As for coffee, Valerie liked hers black and unsweetened, whereas we Jubans enjoy a splash of foamy black espresso in our demitasses overflowing with sugar—Cuba's best export after Jubanas.

"Valerie brought out de choogahr johs for us," Mami says. "Because I told her from day one dat dat wasn't ohpshonal. Eet was, like, 'Here come de Cubans!'—an' reach for de choogahr, honey."

The sugar and the closet. Mami was the best-dressed refugee around, courtesy of the exquisite castoffs from Valerie's perfectly and ongoingly edited yet significant haute couture wardrobe. One dress was more kill-me-now beautiful and luxurious than the next. The flawlessly tailored and lined little black day dress in silk and wool crepe that Mami wore with pearls and killer heels to her mental hospital job interviews? Valerie's old Jackie Kennedy–worthy Nina Ricci. It was like the rhyme I sang with my Baptist divas when we played patty-cake, only taken to a finer level: "Miss Mary Mack-Mack-Mack, all dressed in black-black-black, with silver buttons-buttons-buttons, all down her back-back-back . . ." The orgasmic sleeveless hand-beaded satin and silk brocade ball gown in a pattern of swirling big tomato-red and black and gold bejeweled paisley on an ivory background with the triangle peekaboo cutout between the bosom and the waist, the dreamy confection that trailed along the floor in a rustle like a happy woman's sigh? The festeev dress Mami hardly ever wore because she was a ceeveel servant? Valerie's old Metropolitan Museum of Art Costume Institute–worthy Ben Reig.

Bliss for me to be there inside Valerie's dressing room, luxuriating and lingering in the lushness of superb fabrics—shantung, organza, chiffon, satin, Italian cotton pique, scalloped lace, cashmere, crepe de chine—that embraced me in their fullness from their hangers as I sank into them like a vertical hammock, breathing in my ladies' smoke. T. S. Eliot, it *was* perfume from a dress that made me so digress, digress right into a little girl's projected dreamland of what being a real lady would be like someday. Colorful rows of strappy sandals, mules, slides, and slingback and pump stilettos (lots of spectator styles in there), each pair light as meringues; impossibly thin ivory and black silken slips and half slips that smelled of Joy; sell your firstborn to the Gypsies for the sapphire, pearl, diamond, gold, and platinum pieces of estate jewelry. It was all like being inside a more compact El Encanto. You'd have to be etherized or expired not to *lohvee.*

To imagine having all this taken away, wrested from your clutches by some illiterate asshole hick guerrilla reeking of B.O.—*Gahd!* Valerie lived the way we did before Castro. To never have to think about money or comfort or well-being or peace of mind or miss what used to belong to you . . . Feeling so close to everything we lost . . .

I burst into tears.

"¿Qué pasó?" Mami shrieked, abruptly turning to me, bare-

foot and only half zipped into a fitted sleeveless poufy-skirted black silk Dior cocktail dress with a plunging neckline and a rhinestone buckle on the matching belt encircling the tiny waist.

"Gigi?" Valerie said, her nicotine and alcohol-marinated voice a beacon of eternal reassurance.

"I'm too happy!" I said, standing up. Something hard was digging into my naked archless cutlet. It was the heel of a pointy black silk velvet stiletto with an ornamental rhinestone accent on the vamp, meant to be worn with the dress Mami was trying on. I picked up the shoe and held it in my hand tenderly, as you would do for a delicate baby bird fallen out of its nest or a holy relic from a lost realm.

"I'm too happy," I repeated, sniffling and steaming up my horrendous baby-blue cat-eyes. "God must have sent you to rescue us."

Valerie hoisted me in her long, tanned arms and squeezed me tight to her, kissing my nape. She smelled like expensive flowers. Roses, maybe, with a touch of . . . was it freesia? Lily? Hyacinth? Peony? Lilac? All of those, none of those?

"This child," Valerie said. "I'm wild about this goddamn child!"

My brother Eric's diaper might've needed changing, but that didn't deter him from his crawling rounds at the forty feet of our twenty dinner guests. We could now accommodate more visitors in our home at one time because we'd moved from our "drama an' a reevehr" apartment into a little town house nearby in a development called River Park. Our new place had central air and heat and a sliding glass door leading to a small enclosed backyard with grass in it. Eric and I each had our own room. Mine was upstairs, overlooking the street and directly above the front door. My bedroom ceiling was sloped, which I liked because it lent the room an attic-y feel, and the walls were painted deep, bright aqua, which I loved, green being my favorite color. And the furniture may have been inexpensive—most of the pieces were wicker that had been painted white—but at least it was comfortable and pretty and we'd bought it ourselves. No more of that horrible rented junk.

"He's adooorable," said Josephine from Mississippi, her eyes on Eric. Fiercely entrepreneurial from birth, my little brother had a shoe shine kit and he was going around the room with his wooden box containing brushes, buffing cloths, and cans of black and brown shoe polish. His asking price for a shine was a quarter but people usually tipped him a dollar.

"Tha-yutts just precious!" Josephine added, extending her leg to Eric's ministrations and exhaling her Salem cigarette smoke. "This boy is somethin' else. Go ahead, sugar, have at mah shoes."

Had his forte been, say, sketching quick caricatures, Eric would've done it. The disciplined, driven way he worked rooms was just amazing. His knack for money-making would "take him far," Josephine's husband, Joe Abraham Lincoln, noted. My nice third grade teacher, Mrs. Scott, and her husband thought so, too. (Mami always invited my teachers and their husbands over for dinner.) I'm so sorry, did I miss something? Did somebody have a secret how-to-get-ahead-in-the-world meeting while I was busy poring over my *Seventeen*s and writing my little poems and stories on scraps of paper and playing with my redheaded Barbie in her hot-pink patent leather wardrobe case?

"I want to talk to you," Valerie told me, putting down her cocktail and extinguishing her cigarette in an ashtray Mami had stolen from La Omega, a Cuban restaurant in Adams-Morgan. I climbed into her lap, facing her and careful not to show the other guests my floral *calzoncillos,* panties, under my sleeveless turquoise, violet, and lime faux Pucci cotton micro-mini-dress.

(Unlike Mami, I always gravitated toward happy colors. Life was tough enough living with former mental patients' suicidal black-and-blue self-portraits hanging on every wall, thank you.) On that summer night I also had on a stretchy turquoise hair band that matched my shift. I liked being coordinated. All Jubanas must be. I had accumulated a vast collection of assorted hair bands, barrettes, ponytail holders, and clips; by then my hair was down to my waist. I wrapped my legs around Valerie's hips, tucking the naked cutlets underneath. I lived for nights and weekends, when I could liberate the cutlets from their cookie confinement and be barefoot. Valerie took a good look at me, smoothing the little soft short wisps hovering above my forehead.

"What do you want most in the world?" she asked. *To memorize everything you wear so I can look like you do.* I did have a seemingly useless knack for acute fashion memory. Regarding beautiful women dressed beautifully was instructive and inspirational, not to say natural. I'd done that since I was a baby. Valerie was wearing a navy boat-neck pullover with white stripes and three-quarter sleeves, yellow gold and iridescent white pearl Chanel button earrings, a gold watch with a red alligator strap, white capri slacks, and quilted red satin ballet flats. Very Lulamae–Holly Golightly. Thanks to Valerie, I was learning the lexicon of fashion. She said, "God is in the details."

"What do I want the most?" I said. "No brother."

I rolled my eyes as Eric industriously shined his fifth pair of shoes. It was as if he had on invisible blinders and could see only what was right in front of him. Mami and Papi regarded their son with awe and infinite admiration. You know, Judaic-Latino parentals, firstborn male, etc. I'd been dethroned by a towheaded toddler in wet diapers who shined shoes for money. But! If your parents insist on having sex (whatever that was) without using protection (whatever *that* was) and refuse to consult you so you can, like, weigh in, there's really nothing you can do about it. I knew one thing for sure, though, and nobody had to articulate it to me in words: This was about *princesa* dethronement, Hispanic style, by the arrival of brothers. This isn't a simple sibling rivalry thing—that's kid stuff. This is having to confront the profound Hispanic reality and dirty not-so-little secret that no matter how much your parents love you, they will always prefer their sons. You have eggs. Your brothers have penises. It's as simple as that. It's a general worldview, boys and girls are not equally valued. Hence, not equally treated. No matter how enlightened you may believe you are as a Mundo Latino member, penises beat eggs. Every time. There's no contest in a Latin home. Egg girl, joo lose. Really. We Latinas can't even keep our own eggs as *eggs;* the noun's been co-opted by LatinOs as slang for their testicles. THEIR testicles. Yep. Machismo's not fair, it's not right, but it is what it is. Consider the breast. *Seno.* It's masculine, for God's sake: *El seno.* The butt? Both forms, polite (*fondillo,* or seat) or vulgar (*culo,* or ass)—are also masculine. Okay, so now we've forfeited our eggs, breasts, and butts to *hombres.* What do we have left, anatomically speaking? *La cabeza,* the head (and, accordingly, *la mente,* the mind). As Mami Dearest would say, joos eet or loos eet.

"Besides no brother," Valerie said. "What else do you really want?"

"Normal parents?"

"Besides that."

"I want . . . I want to write," I heard myself say. It was a startling self-discovery. I'd never said it aloud. Giving it a voice made it not just real, but *really* real. For the first time, I'd let someone in on my deepest un-Latina secret. Un-Latina because we're supposed to feel that just loving a man is enough for us. It felt scary to tell Valerie, but I knew that she, a Benign WASP, was the right person to entrust it to. "I have been writing," I added, "but it's all over the place, everywhere."

"What do you mean?" Valerie asked.

I slid off Valerie's lap and tugged her hand to come follow me.

My mother glanced up at us from her conversation.

"We're goin' places," Valerie told her.

Upstairs in my bedroom, with Valerie sitting on my bed, I scrambled around collecting all the scraps I'd used to write on. Mostly they were little sheets from Papi's prescription pads. It had never occurred to me to use my black-and-white speckled school notebooks—purchased, along with all my other school supplies, at Peoples Drug Store or Drug Fair—for anything other than homework. I needed a clear division: personal writing/school-work.

"Oh God," Valerie said, taken aback at the sight of the colony of swarming wasps who'd built a comb in the corner outside my window. They were huge, like monster yellow jackets. Periodically they'd fling themselves with a *ping* on the glass pane, thinking it wasn't there and that they could get me. Sohkehrz. I found the wasps creepily captivating. I'd tap on the window and they'd go *loco.*

"I know," I said, delighted. "They're so gross!"

"But you need to get rid of them," Valerie said. "Wasps are destructive little bastards."

"They don't destroy *me*," I said, sitting next to her and placing all the Rx's on her lap. "I don't wanna hold their hand, though. You like that Beatles song? 'Oh yeah I'll tell you something, doo-doodoodoodoo, I think you'll understand . . .' "

" 'I wanna hold your hand!' " we sang in unison.

" 'And when I touch you I feel happy inside,' " I continued. " 'It's such a feeling that my love I can't hide, I can't hide . . .' "

" *'I can't hiii-de!'* " we sang together loudly, laughing at ourselves. This was so much more fun than the stupid shoe shine party going on downstairs.

"You know what?" she said. "I think what you need is a diary."

"A diary?"

"A place to keep all these in," she said, meaning the paper pile on her lap. There was a poem about wishing for a puppy I'd name Martini (after Valerie's favorite drink), and a map I'd drawn of Cuba with tear-shaped raindrops falling down on it and stick people with upside-down U's for mad, sad frowns. I'd labeled my picture "Castro Cuntri," with a swastika on either side of the title.

"A diary is just what's needed here," Valerie said. "You can put it all in one safe place. Then you won't lose anything precious."

"Anne Frank! Like in the secret annex up in this room. And the wasps can be like the Nazi bullies outside, the S.S."

"Yeah," Valerie said. "Only you, *señorita* in a Pucci . . ."

"*Faux* Pucci."

"You will have a happy ending."

"*Dios,*" I said, "is in the details."

"Right."

I jumped on Valerie so hard to hug her in my jubilation that I knocked both of us back on the bed. I hadn't felt this good since that moment in the Varadero sea when something beautiful in my body vibrated, awakening me.

CHAPTER FIVE

Finding My Black Parents

*y Jewish friend Holly was pale and slender and quiet, with very dark straight hair and thick bangs, dark eyes, and full lips. Holly's chain-smoking, deep-throated, red-headed New York City mom was Louise. Louise ran a drama school called Stage Studio on Connecticut Avenue at Dupont Circle in Northwest Washington. Sharing Red Hots, Smarties, and Lemonheads after school, Holly would tell me stories about her mom's rip-roaring days in New York, when she was an actress and something called "a hoofer." I liked the sound of it. New York, I imagined, was an island as openminded and exotic and sophisticated as Cuba. Mami talked to Louise, and soon I was enrolled in Saturday afternoon drama class, right after my morning ballet.

But liking the sound of another's life and wanting to get into it yourself are two different things. The only reason I couldn't go into hiding under my bed for drama, *a new experience and therefore terrifying*, was logistical. I'd finish my two hours of ballet, Mami would pick me up, and we'd drive straight to Dupont Circle.

"Don' worry," said Mami, aggressively tailgating the sohkehr going too slow for her taste in front of us. "Jool lohvee."

"Like I have a choice," I said. "Couldn't we at least have gone home to change? I'm in a leotard and ballet slippers, for God's sake. My hair's in a chignon. They'll think I'm retarded."

"Joo know what? Eef dey do, deyr de retardeds, honey. Tell dem, 'Fohk joo! I am a *bailarina* as well as an actress an' JOO are johs a notheengh one-deemehnshohn zero weeth notheengh else to offer een all de life.' "

"I'll quote you."

"Please do."

I didn't mean to the students. I meant in my diary. Valerie had given me a dreamy one (which, if it wasn't lost when we later moved to Silver Spring, Maryland, is probably tucked away somewhere in my parents' house): a stiff-backed journal covered in silk cloth in a plaid pattern of predominantly pink with black, turquoise-blue, green, yellow, violet, and red on a white background. The heavy vellum pages were all lined, and there was a brass lock on the strap, with its very own tiny old-fashioned brass key. Though I always wanted to keep some kind of record of my life, I initially hesitated defacing this treasure with my little scribbles. But Valerie said, "Oh hell, mess it up. All you want. That's what it's there for. Writing is a messy business! All the greatest writers throughout history have been messy people, everybody knows that."

"What happens when I fill it up?"

"Call me."

To my Baptist divas' parents, however, I would quote not Mami but Martin Luther King Jr. or Langston Hughes, figuring the latter's patois would be more appropriate for that given audience.

(I'd memorized Hughes's famous poem, printed on the first page of *A Raisin in the Sun,* a copy of which Alfie Brown had given me.) Here's why: The tradition at Amidon Elementary was that whenever there were birthdays, the birthday child's parents would provide cupcakes for the entire class. When it was my turn, Mami and I would go to the Safeway or Giant grocery store bakery and order a couple dozen to bring in. But kids had birthday parties at home, too; it's important to keep the sugar going and the presents flowing. So I'd be invited to my little friends' galas, and Mami and I would go out to buy me birthday clothes at Best & Co. on Wisconsin Avenue in Northwest, across the street from Sidwell Friends' upper school campus. Mami picked out a dress with a dropped waist, fitted elbow-length sleeves, and a full skirt—*tafetán color champán,* of course. (Never too early to get your Jubana child into the ultimate, aka marital, groove.) We'd go to Rich's Shoes, farther up Wisconsin Avenue in Chevy Chase, over by Saks Fifth Avenue, where one of Mami's former patients was a salesman. I tried on thin-soled maroon patent leather Mary Janes with squared-off toes and a single pearl button clasp on the outer end of the thin straps that traversed the cutlets. The shoes were sleek and slim, obviously meant for a narrow-footed *princesa,* which I was not.

"They're pretty but they hurt," I said, looking down as my medium-width feet created a bulge where there shouldn't be, thereby ruining both the line of the shoes as well as my God-given right to have shoes that fit properly.

"Hurt?" Mami said. "Hurt where?"

"Everywhere," I told her. "They're too tight."

"No. Deyr not. Deyr great! Ees de right size."

"Bitch, too tight *here,*" I said, touching the bone that protruded just below my big toe.

"What ees joor point?" Mami said, taking out her wallet. "Joo have de same feet as me. Narrow. All the Jubanas *mas finas* [most refined] have dos same feet. Jackie Fohkeengh Kennedy has dos feet, okay? An' her husband ees dead!"

"I may have to kill myself," I said. "Or you."

"Chee looohvs dem," Mami told her former patient. "We'll take eet! An' also dees an' dees an' dees, too. Thank joo SO much."

There were stacks of open shoe boxes all around us. No way any self-respecting Jubana exerts the quaint Anglo concepts of "self-control" or "restraint" when confronted with more than one pair of beautiful shoes—or beautiful anything else, for that matter. In our world, *mas es mas.* More is more.

"Somehow," I said, picking up a gray suede Hush Puppy loafer with black crepe soles and swirling black tooling on top, "I just don't hear these calling my name."

"Dos are for de school days," Mami said, signing the receipt with her distinctive signature small, round, fat script and hollow circles for dotting I's and making periods. This style used to drive her Cuban professors crazy, the way being left-handed used to be considered bad or wrong in the old days. But as usual Mami persisted, and I learned to punctuate the same way myself.

"I'd rather have black leather riding boots than these gray things. When am I ever ever ever ever EVER going to get those? I've told you a thousand times."

"Joo know when? When joo get joor own leetl horsee."

"And when will that be, exactly?"

"When joo get joor own leetl job to pay for joor own leetl horsee an' eets own leetl barn an' de leetl black rideengh boots."

"I'm in *elementary school,*" I reminded her. She evidently required a reminder of my world status. "It's illegal for me to work!"

"Das joor eemeegration problem, honey, not mine."

"So glad we cleared that up."

"Me too. Now ees time to go an' smoke an' have a leetl drin-kee. I need to relax. Chohpeengh ees exhausteengh. Motherhood ees a beech."

" 'Shopping is exhausting.' Please. It's your favorite thing in life."

"Dat ees so wrong. My favoreet theengh een de life is chohpeengh . . . a LOT."

On the day of the given birthday party we'd get me all dressed up—I initially played along to divert Mami—and there I'd be in my *tafetán color champán* dress, opaque ivory tights, patent leather maroon flats, horrendous baby-blue cat-eyes, and long hair pulled back into a fragrant Agua de Violetas–infused ponytail, with a strip of ivory satin ribbon tied in a bow to cover the band. Mami would have the girl's gift all nicely wrapped—and I'd dive straight under the bed. Every time. Then we'd commence the same routine: Mami cursing and grabbing a stray foot or arm and dragging me into the car, the two of us yelling at each other all the way to the kid's house, Mami slamming on the breaks for pissed-off emphasis (nearly hurling us both straight into the windshield), walking up to the door, plastering abrupt fake smiles on our faces, Mami ramming me inside the house and saying buh-bye.

It would take me a few awkward minutes to acclimate to the social environs. And then I'd be fine. More than fine. After the obligatory time with the kiddies, I'd go find the more (for me) socially suitable parents, and proceed to endear myself to them.

"I looove your house!" I'd commence. "You must be really happy and proud that you live here."

And they'd say in an aw-shucks kind of way, "Well, yes. Thank you!"

Then depending on the size and looks of the place, there'd be two alternate responses. Either (1) "You are so welcome. And I notice that there's a lot of room. So lucky. We live four to a room because of our Cuban immigration problems. It's kind of hard. Black Americans have had to face so many similar things." Or (2) "You are so welcome. And I notice how COZY it is here. Sooo cozy. It reminds me—I'm sorry, I don't mean to cry but it's so hard when I think back to this—of our beautiful tiny little home back in Cuba. You know, it wasn't much, but just having the family all together was so wonderful. And now . . ."

Either way I'd be smothered in parental hugs and kisses and given extra helpings of coconut cake and lemonade, two American items I adored and never had at home. As with any ice cream flavors but coffee and pistachio, whatever taste didn't compute for Mami, we simply never ate. Lemonade? Mami looked down on it. Cubans do limes, *gracias*, never lemons. Coconut cake? The closest I ever got to that at home was *flan de coco*, coconut caramel custard, which Mami made the hard-core Cuban way, with canned *coco rallado en almíbar* (grated coconut in heavy syrup), three hundred egg yolks, and a can each of condensed and evaporated milk. Hey, we started with the *café con leche* laced with eighty-seven tablespoons of *azúcar* at birth; this is nothing for us. *Dulces, pasteles,* and *postres*—sweets, pastries, and desserts—are not our island's most subtle or airy delicacies, not that Cubans have much subtlety to offer the world. As poor dead Pauline Kael said of movies, "It's got to be too much or it's not enough." She could have been Cuban, actually. Consider our beloved *tres leches,* three-milk cake, which we co-opted from the Nicaraguans. God forbid our cakes should have just one milk, that's for anorexic amateurs or crybabies. Our sweet shortcake is soaked and layered in evaporated milk, condensed milk, and heavy cream. (The latest Miami version is *cuatro leches,* a *tres leches* with the addition of *dulce de leche,* a caramel sauce made of sugar, cream, and butter.) Or try some *turrón.* Calling the traditional holiday treat "a nutty Spanish nougat candy" is like calling

Middle East politics "untidy." Either way, this decadent imported goodie, whose heavy reliance on ground almonds is a Moorish-Arab influence, is not for the faint of heart or diabetes-prone. The two main kinds of *turrón*, and both come in rectangular seven-ounce bars or bricks, are *jijona* (smoothly textured, hyper-rich, mega-sweet, oily, with a gritty trace of ground nuts) and *alicante* (like a stone wall of almond-studded nougat, less sweet than *jijona*, it requires a hammer and Phillips screwdriver to break into edible pieces). Cubans are really passionate about their favorite *turrón;* I myself am partial to the *jijona*. One time I bit into a jaggedly hacked *alicante* chunk and it fractured a molar.

So there I sat on the birthday girl's mom's or dad's lap, happily consuming my lovely light coconut cake and lemonade, with the damn party shoes kicked off the now aching, throbbing cutlets. Hey, no birthday girl ever minded that I hogged her parents. She was grateful I got them off her back so she could go off and play and do whatever she wanted. But I, an alienated and deprived political refugee child in search of Other Parents, obviously had much more important work to do than these immature, monolingual American kiddies did. So I got to it. After all, the parental grass is always greener on the other side, no matter what skin colors are involved. (I'd have glommed on to Valerie, my first choice naturally, but she lived much farther away from us than these here parents, so she wasn't as accessible. As for Latino families, there weren't any around, besides mine. You have to be realistic when you're trying to switch parents.)

"I love you *so* much," I'd say. "I love you so much that I have a dream. Yes, I have a dream today. Want me to tell you what it is?"

Naturally they'd say yes.

" 'The Negro lives on a lonely island of poverty in the midst of

a vast ocean of material prosperity,' " I'd begin, reciting key portions of Martin Luther King Jr.'s famous speech, the one he gave
on August 28, the hot summer day before Eric, my business
prodigy brother, was born to shine shoes and eventually make a
fortune in pigdom. We'd talked about King in school, and my
teacher had handed out mimeographed copies of his entire address, certain portions of which I'd memorized, as I felt they spoke
directly to me, an honorary Negro love child. Besides, in acting all
you do is memorize other people's words. This was a cinch. " 'The
Negro is still languished in the corners of American society and
finds himself an exile . . . And so we've come here today to drama-
tize a shameful condition.' "

I'd wait for a reaction. Usually it was wordless wonder. Then
laughter.

Same thing happened when I walked into that drama class for the
first time. Holly's mother, Louise, introduced me to the class. She
was a no-nonsense dame, with pockmarked skin, a cigarette
hack, and a tough demeanor. She was nice to me. She felt sorry
for me and my refugee straits. I scanned the faces scanning me.
Not another prepubescent in sight. (Holly never took drama, and
Louise felt I'd be better suited to being placed in an adult class
due to my "precocious social maturity.") Then, just as I'd pre-
dicted to Mami, they all laughed at my ballet clothes and ballet
hair and heinous baby-blue cat-eyes. Louise told them to knock it
off, that I was a Cuban ballerina as well as an aspiring actress.
Truly, I didn't see myself as either. The dancing and the drama
were more for Mami, a vicarious way for her to overcompensate
for my tragic deficits. I just wanted to get back to my room and to
my beautiful new diary to fill it all up and to call Valerie like she
told me to and see what would happen next.

In class we began with the basics—sense memory, action, beat, business, character, improvisation, objective, obstacle— and eventually I got my first real part, that of Helen Keller in *The Miracle Worker,* a play that allows mostly white people in it. The role was beyond perfect for me, as in it I could remove the glasses and use my acute myopia to my advantage. Plus there's a scene or two where Helen gets to *act out*—another term I'd picked up back at St. Elizabeths—by physically attacking Annie, her teacher. This was one of my fave moments because being loud and theatrical is so utterly Cuban.

Armed with my newly acquired theatrical skills, I would soon realize my covert agenda to get black parents. Back on the parents' laps at the birthday party, I pressed on. *Action, beat, business, character, improvisation, objective, obstacle.*

"Well, I'm like a Negro because I come from a lonely island, too," I continued. "Only it didn't used to be lonely but now it is lonely and so are we, lonely like homesick, because of Fidel Castro, Hitler's cruel demon spawn. And Cubans, now we're in *el exilio.* Like our own poor little island right inside the richest ocean, that's the United States. All because of that wicked spawn. I'm gonna stick needles and pins into him. Long sharp ones for when he's dead. And then we're gonna get our house back and our country back and my air-conditioning with the painted baby bears on my dresser and then we'll be very happy again and I know we'll go swimming in the sea every day with the *caracoles.* Seashells. You can come, too. I'll show you it. We can even have a little siesta after lunch."

I'd catch my breath for a beat and dive back in.

"So then recently, on my seventh birthday, December 10, I saw and my mommy showed me how Dr. Martin Luther King Jr., he got the Nobel Peace Prize, you know? It was on a Thursday, I'll never forget it because that was my birthday! Dr. *Rey*—that means king—he said that peace costs more than diamonds or silver or gold. Well, my mommy just drives me crazy! That's not peace! So anyway I was wondering about if you could think about adopting me maybe. Just until the evil *asesino* croaks and then it's time to go back home."

My audience of two would crack up and hug me and kiss me and tell me all sorts of nice things about myself. That was the good part. But then they always added that my parents loved me and I belonged with them. Now, another Jubana might have felt defeated at that point and given up. But another Jubana might not have anticipated this. I didn't fall off the *turrón* truck yesterday! I could out-manipulate. I'd learned that at the scrawny neck of my *gringa* school principal and even more so by playing Helen Keller, a master manipulator if there ever was one (until she got the tuf luv and the gift of self-expression). So. I was not prepared to give up that easily. The resistant black parents, in other words, had forced me to resort to playing the race card.

"Excuse me, but are you saying that you can't adopt me because . . . I'm . . . *white?* Is this, like, a racial thing? Because I know I could belong here. I couldn't be in *A Raisin in the Sun,* Alfie Brown said so, because it's a black folks' play. I have to be a white blind girl whose family doesn't understand her. It takes an outsider to. *The Miracle Worker.*"

Then for my finale, I'd launch into my MLK–Langston Hughes combo patois platter.

"I *told* you, I have a dream today. 'What happens to a dream deferred? Does it dry up like a raisin in the sun? Maybe it just sags like a heavy load. Or does it *explode?*' "

By the time Mami arrived to pick me up, my potentially adoptive black parents regarded her exactly the way I wanted them to: monster child molester.

My dream of adoptive black parents never did come true, parents I imagined would give me infinite love, affection, under-standing, warmth, protection, support, limits, and time, not to mention coconut cake and lemonade. Yes, it's true that I ulti-mately failed to get black parents. But I feel I came close, and the effort was in itself rewarding.

That's something, I think.

Some dreams do not die, especially when they involve Valerie.

When I had filled up my diary, I called Valerie like she said to, and the next thing I knew I had an Hermès typewriter, all green, with a hard dark green lizard cover with a brass snap on it. The typewriter wasn't just beautiful; it was a real workhorse made for pounding. I'd begun teaching myself how to type. Since I'm right-handed and my strongest finger is my right index finger, I used it exclusively. I was gradually working my way up to big speed. If all you did was listen, you'd swear it sounded as though I was using all ten fingers. Valerie had also given me two ribbon refills—with double black and red ink bands—and a bottle of Liquid Paper, a whole box of white typing paper, a stack of file folders with labels, and a pack of brightly colored indelible Magic Markers. I was ap-prehending a crucial North American lesson: You can't just dream dreams. You need the tools. This is why there's Home Depot. Staples, too. But they didn't have those back then. As Mami always told me, *"Allí tú."*

Joor on joor own.

CHAPTER SIX

Always Wear Waterproof Mascara

*S*earching for new parents as a Cuban refugee child is a bitch. I am not jooneek. You know it, you saw it, as I did that lazy Sunday in mid-April of 2000. The fiancé, who was my then-boyfriend, and I, both journalists, were eating bagels with cream cheese and drinking *café con leche* as we read the *New York Times*. Normally we'd be listening to American jazz or a Cachao *son*—a lively, sinuous, romantic, rhythmic style of Cuban music—with the TV turned off. But this was not a normal day or a normal spring. Elián González, an adorable six-year-old Cuban refugee, was obsessing me and, so it seemed, the rest of the world. By now his story has entered into Myth Land, but at the time, it was a compelling myth-in-the-making, a heartbreaking political and emotional saga, and a daily breaking news event with no clear-cut outcome that could ever satisfy everybody except, ultimately and as usual, El Caballo, Fidel Castro, Hitler's demon spawn.

To quickly recap the facts: Elián, whose parents were divorced, lived with his mother and stepfather in Cárdenas, a port city two hours east of Havana on Cuba's northern coast. On November 22, 1999, the three Cubans joined ten others in a sixteen-foot motorboat heading for the United States. The next day, the vessel capsized, drowning ten of its passengers, including Elián's mother and stepfather. After clinging to an inner tube by himself for some seventy-two hours in the Gulf Stream in the Straits of Florida, Elián was rescued near Fort Lauderdale on Thanksgiving Day by two American fishermen, who delivered him to some distant relatives living in Little Havana, a conservative working-class Miami neighborhood of Cuban exiles. (The two other shipwreck survivors, both adults not related to Elián, separately got ashore, but nobody cared about them.) Elián enjoyed a tearful, joyful union with his extended family, who intended to keep him.

Then the proverbial *mierda* hit the fan.

Castro and Elián's father, Juan Miguel González, demanded Elián be returned. Predictably, the Miami relatives refused. Juan Miguel, a member of Castro's Communist Party, arrived in Miami to reclaim his son and return him to Cuba. Cuban Miami exploded in a collective frenzy, turning an innocent child into a political pawn in an international custodial tug-of-war, a Christ-like symbol of the miraculous and the preordained, and a poster child for all Cubans.

"Is there any more coffee, Geeg?" asked the preengagement boyfriend, Paul.

I was too engrossed in the TV news show to hear him. Talking heads were butting heads over Elián.

"Geeg?" Paul repeated.

"What?"

"Is there more coffee?"

"Here," I told him. "You can finish mine. Sweetie, I really want to watch this."

Paul understood, and not just because he's a journalist. My

gringo boyfriend of one year, looking mighty cute in his Romeo y Julieta–brand cigar T-shirt (I'd bought it for him many months before in a Little Havana *tabaquería,* cigar factory, I wrote about for the *Washington Post),* had by now gotten a deep immersion in all things Cuban, thanks to me and my family. Although my parents and I didn't remain in Miami Land after leaving Cuba, there were always summertime trips to Miami, the Other Homeland, to visit friends and relatives and overeat at Versailles, my favorite Little Havana restaurant. Those vacations became increasingly less fun for me as time went by, however. The eternal, crushing heat. The political tunnel vision and dogma. The vanity and materialism and concern with appearances, with *¿el qué dirán?,* the what will they say? The nonstop Castro-bashing mania over the airwaves, in restaurants, at people's houses. The clannish, provincial hysterics. All those things would have been my norm had my parents settled there instead of in our nation's capital. And yet, sometimes I've been sorry they didn't. After all, there's strength in numbers, a sense of belonging and acceptance and solidarity.

In Miami, Cubans know who they are. They never have to feel ashamed to be it. That's one of the things I love most about going there, that instant unspoken understanding that *nosotros somos Cubanos.* When you're such a minority all the time, it's wonderful to be a part of a majority for a change. On the other hand, most Jubanos are Democrats, which puts us in the minority among overwhelmingly Catholic Republican Cubanos. Considering that all Cuban Americans are the tiniest Hispanic group in the United States—an underwhelming 3.7 percent, or about the same number as Jewish Americans—Jubanos are *really* in the minority. (According to the 2002 Census, most U.S. Hispanics are Mexican, 66.9 percent; then Central and South American, 14.3 percent; Puerto Rican, 8.6 percent; Cuban, 3.7 percent; and the rest make up 6.5 percent.) Don't even get me started about the blank, polite, Thorazine smiles on Jewish American faces when you tell them you're a *Cuban* Jew. Pass the halvah and pass the pork. At any rate, when I'm sitting in Versailles, the weirdos are the Americans who can't pronounce *plátanos maduros.*

So although I obviously feel very Cuban most of the time, especially in contrast to my New Yorker fiancé and his North American family and most of the rest of the native English speakers in my northeastern life, I will nevertheless never feel as fanatically or defiantly Cuban, or as Cuban to the exclusion of all other things, as do my friends and family who left Cuba in the early sixties, moved to Miami, and haven't budged since. Still, when a crisis arises in the community—and the fate of little Elián was certainly considered *primo* crisis material—we tend to circle the wagons.

Since Paul had polished off the remnants of my *café con leche* and I needed more caffeination to couple with my freshly lit Parliament for my Elián fixation, I unscrewed a fresh bottle of TaB. Yes, bottle. TaB is available in most metro areas, but usually only in cans. Canned TaB tastes metallic. In a pinch, I can settle, but for some bizarre reason, only TaB in the fabulous twenty-ounce plastic screw-top bottles works for me. Sometimes I have to drive more than two hundred miles one way just to get my hot pink cases of those TaBs. You should see them piled up in my trunk and backseat. Think of my shock absorbers! And then having to haul them all out and bring them upstairs into my apartment and find a nonobtrusive spot for forty cases of TaB. And then people come over and see the wall o' TaBs, and I have to pretend to laugh it off and say something like, "I know. It's an ironic statement. Neo-Warholesque, if you will." All because I have an addiction. Paul said he knew I was Really in Love with him the first time I willingly shared my TaB. Well, what can I tell you? A girl goes *puerco* wild when she's in love. Wild and blind. Just as my fellow Cubano Americanos on TV were acting over Elián, a boy whom the majority didn't personally know but with whom they were madly in love, madly to the point of blindness.

Watching the live footage of my fellow exiles in furious full-throttle right-wing mode on TV, threatening to sue attorney general Janet Reno and the U.S. government, and weeks later erupting onto the streets of Miami when Elián was finally returned to Cuba with his father, I was truly torn between cringing (Americana: God, these freaks are so embarrassing!) and empathizing (Cubana: We may be short but we're fierce. You go, kids!).

So. Which camp would I side with, which one should I side with? But then I thought of a lesson I learned from Dr. Marvin L. Adland, my retired psychoanalyst-expander, and an exceptional student of the human condition. It was one of the hardest lessons to get locked into place inside me. Namely, you *don't* have to raise your voice or use bad words to be effective and get your point across—which is, of course, the ultimate anti-Cuban attitude. As a matter of fact, acting "Cuban" in that way in certain milieus of this society actually weakens people's perceptions of you and makes them think you're just . . . a typical, trivial, crazy, hot-blooded Spic.

Maybe we exiles drink too much espresso and we're wired from the caffeine. Maybe we live in chronic sugar seizure mode from our ultra-rich sweets like *flan de coco* or *pastel de tres leches* or *turrón.* Maybe we just have too much downtime. (Indeed, ask any Cuban old enough to have lived in pre-Castro Cuba about life in pre-Castro Cuba and you'll be incredibly sorry you ever brought it up. Ask about post-Castro Cuba and you'll be even sorrier.) I think what we really need isn't less caffeine and sugar and free time; it's better public relations. Because if the Miami crowd's goal was to make mainstream America feel sympathy for the cause and understand that Castro's repressive, anachronistic regime was behind all of this from day one—dey totally blew eet, as my testy Mama Jubana would say.

A few days earlier my mother and I had been on the phone, both of us watching the same CNN show in which "famous" Cuban Americans had formed a human chain that they called a prayer circle in front of the Little Havana house where Elián was. There was a well-known actor, a talk-show hostess, a singer, a musician, a mayor, and, to my personal astonishment, a newspaper colleague whom I knew fairly well. I could perfectly understand the celebs getting out there—all publicity is good publicity, after all—but a *journalist?*

"If I were her editor I'd fire her Cuban ass on the spot," I heard myself tell Mami. I was surprised to hear myself pop that out just like that. "She's entitled to her opinion—in print. But she's injecting herself into the news. That's not journalism. That's—"

"De newspaper might talk to her but dey won' fire her," Mami said. "Chee has a consteetuahncee. De paper knows dat. Der would be an uproar. Dey have to take eet. Chee knows exactly how far chee can poosh her agenda."

Back on the Sam and Cokie show on TV, George Will was saying, without a trace of irony, that a Communist cannot be a good parent.

"There's a howler," Paul said.

Was it a howler statement? Or was this situation a Cuban exile version of "it's a black thing, you wouldn't understand"?

"Jesus Christ," Paul continued, "Will's such an anal-retentive Tory."

"You think so?" I said, lighting a new Parliament and sipping my TaB.

"You know what the Justice Department should do? Reno should get a court order that the loony uncle [Lázaro González, at whose home Elián resided while in the United States] cough up the kid. And if he refuses, they should lock him up for contempt."

"You would LOCK HIM UP?"

"Absolutely. 'Communists can't be good parents.' Christ."

"Maybe they can't," I ventured.

"Fidel is a hypocritical thug. But the Elián crowd, those people down there are loud, bombastic, right-wing, hard-liner assholes who are trying to use their political clout and campaign cash to put themselves above the law."

Paul picked up my bottle of TaB and took a long swallow. Now not only was he insulting my people but he had the audacity to drink my TaB right afterward, without so much as waiting for me to offer him some. This is my LIFE, dammit. Don't you get that I saw myself in little Elián? It's the story of so many Cubans.

Me cago en este cabrón jodedero! Shit! I was thinking very bad things about Paul, all because he insulted My People and took My TaB. Could I continue to voluntarily sleep with a man who drinks My TaB *and* has an anti–Cuban exile attitude? I mean, it's one thing for ME to criticize them, it's quite another for outsiders to.

"I know what you want me to say, dear," I said, trying my best to not repeatedly stab my *gringo* beloved with the now empty— EMPTY!—TaB bottle. "Like in those multiple-choice quizzes in *Cosmo,* when you know what the 'right' answer is and you choose it just so you'll get a better score and not have to face the fact that your attitude's all wrong and you're all fucked up?"

"What does *that* mean?"

Where would I even begin? How to explain to an American, a non-Cuban, the passions behind this, the frustrations and hurt feelings of more than four pent-up decades . . . ? *Ay, Cuba.* See, this is why I always wear waterproof mascara. You just never know when life will make you cry. What really drives me crazy is how tears uncurl your lashes—but not evenly. Half will be up and half will be down, and all that effort you put into curling them all perfectly with your Shu Uemura eyelash curler in the morning just gets shot to shit. Hate that.

"Elián's mother *died* to get him here," I said, blinking and feeling my eyes brim. "What do you not understand? She wanted him HERE. You know how many other Cuban parents have made similar sacrifices? Hello, there's a REASON for this. I realize no six-year-old gets to call the shots, but . . ."

Coño. Elián was affecting me way more than I'd expected. I'd just mentally cursed out the man I loved. At least I'd kept my volume in check. That's a real accomplishment for a Cuban. Here's the thing that clouded my judgment and got me so riled up: Seeing that terminally cute (which is how Paul describes *me,* actually) Elián swinging on the swings in his Little Havana yard, playing with his five bazillion Toys "R" Us toys, drinking his little *mamey* nectar, and enjoying all the fruits of *Los Estados Unidos,* I was very moved. I identified with him. My parents didn't want me to live in Cuba under Castro, either. They also took risks to get me from there to here when I was a tiny child. So to see all of Elián's mom's effort and risk and even death come to nothing for her son except a U-turn ticket back to Cuba after having lived it up here was just sadder than any words. Who knows what might have become of me had my parents been living apart, as Elián's were back in Cuba, and my mother had drowned trying to get me to America and my father had come here to reclaim me and return me to Castro's Cuba? I'd never have met Paul, that's for sure, or experienced the life-altering thrill that only fashion magazines and TaB in the bottle can bring.

So Paul should be glad and Paul should agree!

Paul was glad—for me. But no, Paul didn't agree—about Elián. Hm. Maybe I should be involved with a Cuban, not a *gringo.* The machismo would turn me homicidal, true, but at least we'd agree on Elián and the importance of *flan de coco, pastel de tres leches,* and *turrón.*

"Geeg?" Paul asked.

"No sex for joo tonight, Mr. *Señor Jahnkee imperialista.*"

"What? Why?!?"

"I'm too confused," I said. "I might be sleeping with the enemy."

"Come *on.*"

"You should have seen this coming, dear," I said. "You know how I feel. You should have either agreed with me or just humored me and gone the hell along with it."

"Why? I don't patronize you. That's ridiculous. We can't disagree about a six-year-old Cuban and move on?"

It was a good question.

"You know what?" I said, lighting another Parliament, reaching for a new TaB, and pulling back in my chair to physically extend his proximity to my elixir. "My *pussay* is Cuban and your 'apprentice' is American."

"Oh my God."

"No more invasions for a while. Okay?"

Know how I was able to be so Cubanly ballsy with Paul, who can be quite ballsy himself when the occasion calls for it? The same way I'm able to be it with anybody who pushes me beyond my pathologically delightful Cuban limits. As Dr. Adland always says, "The only way to have a good relationship with anybody is to be prepared to lose it." (Which is different from Mami's joos eet or loos eet.) Cuban exiles are not prepared to lose their relationship with Cuba and are therefore doomed to go ballistic over Fidel over and over and over until they croak—or he does. I told Paul I needed to walk it off, and he said that sounded like a pretty good idea. I took my cigarettes, lighter, TaB, and keys, and went outside. It's a quiet residential neighborhood in a fancy burb, lots of old apartment buildings, grass and trees. It had rained overnight, and the sidewalks were still damp. As I walked down the tree-lined street, a few stray raindrops falling from the leaves touched my face.

Who do I want to be? I wondered, blinking a raindrop off my eyelashes. (See? You not only never know when life will make you cry but you also never know when it might rain. Waterproof mascara at all times.) Do I want to be a person who dedicates her life to reacting to the arbitrary exhortations of one singularly sicko dictator named Fidel Castro? That's like being an abused wife, no? An abused wife who takes it and complains about it but who can never leave the abuser? My great Puerto Rican psychoanalyst friend Manny Roman would call that the "Let me tell you what he did to me this week" school of victimized female patient-hood. What was my relationship with Castro, really? *Was* there a relationship, except in my head and in the foment of the Miami crowd? Well, maybe indirectly. Oh please. Even I can't make myself believe that. The truth is that there *is* no relationship. Castro is a fact of life but I have no control over him. I can, however, control how I deal with me.

Oh.

I had to walk away from all the noise—the ever buzzing TV set, the Internet, the newspapers, magazines, radio—that had gotten me so wrapped up, caught up, and carried away every day from far away for months, to gain some perspective. Kind of like my family and I leaving Miami for points farther north. Kind of like turning off the news and going for a walk.

The street was tranquil. I saw a faint sun hovering behind a cluster of pale gray clouds, trying to come out. Cars went by, open windows full of music. A passerby smiled. A woman jogged with a dog on a leash. A trio of kids laughed as some raindrops fell on them from the leaves. I had a relationship with Paul, not with Fidel, and I didn't want to lose it. I had wanted to make Paul understand me better as a Cubana, and to persuade him to respect and approve my exile-centric point of view. But now I just felt silly about it because . . . oh God, dare I say it . . . Paul was right. As long as the parents are good, their children belong with them. Period. (Except for me, who deserves to have good black parents. Hey, it's not as strange as it sounds. My godmother, Nisia, is black, and if my parents had died in Cuba, I'd still be living there. Remember, Nisia stayed behind by choice. I'd have been a myopic commie dressed in olive drab fatigues and Ché Guevara T-shirts. Which, come to think of it, is what the adult fashion cognoscenti don as "style statements," poor slobs. I guess it was good that the parentals didn't die and we got the fuck out of there.)

But getting back to Elián, who are we as Cubans, or as anybody else, for that matter, to say what's best for Elián? It's outrageous and arrogant—two adjectives Castro hurls across the water at Miami exiles—to think otherwise.

I lit a Parliament and thought, really thought. I sipped some TaB and considered Paul. I'd had enough hours of expensive psychotherapy and watched too many Dr. Phil shows to know that if Paul and I had broken up over Elián, it wouldn't have really been about Elián; it couldn't have been. Just as when couples divorce over "money issues" you know it's not really about money per se at all, but rather about what money represents, the metaphor of money. There's something else going on, and it's usually a power struggle. Once he's acknowledged and diagnosed that, Dr. Phil always says, "So do y'all want to be right or do y'all want to be happy?"

I put out my cigarette, finished the TaB, and thought about how you can love people, feel tremendous empathy and affection for them, and still not be able or want to live with them. This wasn't good or bad, it was just the way it was for me vis-à-vis Miami Cubans.

I went home. Paul was napping on the sofa, his back turned outward. My gorgeous little cat Lilly was curled at his feet, as sweet as could be. I put down my stuff and noticed the *New York Times* haphazardly strewn across the dining room table, read and discarded. By tomorrow it would be yesterday's recycled news. The TV had been turned off and in its place a Cuban CD softly played an old mambo. My copy of *A Midsummer Night's Dream* was right where I'd left it on the coffee table. I thought about what I'd like to have for dinner; there was still some fabulous *puerco asado*, roasted pork loin, I'd made the night before, and some white rice and black beans and fried ripe plantains, some of Paul's adopted favorite foods.

I walked over to the sofa, knelt down, petted Lilly, and stroked Paul's hair.

"The course of true love never did run smooth," I whispered into his ear, quoting a line from Shakespeare's lovely play. I meant my true love for Paul and I meant the extreme love of Cuban exiles for Cuba.

"Hm?" Paul murmured, moving a little. He was very sleepy.

I traced a map of Cuba on his back with my fingertip, the bumpy brown mole where Havana, my birthplace, would be. There. Now the place and the person I loved most in the world, loved madly, seethingly, and without cool reason were one.

"I love you," I said into his spine, laying my cheek on the Cuban capital. Paul turned toward me, half asleep.

"Love you, too," he mumbled. "Still pissed?"

"Not really," I said.

We kissed.

"It's an old wound," I said, climbing on top of him and laying my body down across his. "It's the whole Cuban thing."

"*You're* quite a Cuban thing," he said, smiling. "My terminally cute little Jubana princess."

I rested my head on his shoulder blade, matching his breathing, smelling the soft cotton of his cigar T-shirt. Unlike my lost island home, my vanished life, and one little shipwreck survivor heading back to that mythic home, Paul was here, warm and alive and real. He was what I could truly embrace.

So I did.

CHAPTER SEVEN

A Marzipan in the Second Act

*C*ould there be a more unfortunate, anti-Jubana color combo than maroon and gray? Maroon and gray were the official Sidwell Friends School colors. I was a turquoise-violet (faux) Pucci gal who would soon discover the thrill, sensation, and rightness of coral and tomato-red lipsticks. I was also a nine-year-old public school fourth grader being interviewed for possible admission into the toniest, most competitive, most elite, most overwhelmingly Anglo-albino private prep school in the most powerful city in the world.

At 9 A.M. sharp I was seated on a maroon sofa in the admissions office in a stone cottage called Zartman House on the grassy fourteen-acre campus. It smelled smart and rich and old, and musty, like faded, dusty Persian carpets and ancient, mildewed books. My interviewer was the middle school principal, Mr. John F. Arnold (B.S., Washington and Lee University; Yale University; University of Houston). He was a Southerner, I believe, an un-Valerie kind of WASP with big pink ears that stuck out like Dumbo's. Mami had taken me to see the animated movie, and she said that elephants are supposed to have excellent memory. So I knew this blond pachyderm was mentally recording every word I uttered and would never forget anything I said.

Not that I was worried about it. I've always aced interviews. Remember the skinny *gringa* principal with the pearls back in kindergarten? Please. Cubans are congenitally charming. We're social *mariposas*, butterflies. Besides, Sidwell was the fourth private school I'd been to visit in that spring of 1967, and so, to answer Jimi Hendrix, yes, I was *experienced.* The school board in my Southwest neighborhood had elected to enact the Tri-School Plan, basically a forced reverse integration program involving my school, Amidon, and two other inferior local public schools, Syphax and Bowen. My parents rightly did not like the idea and were by then in a position to afford full tuition for virtually any private school for me. Their D.C. friends suggested four top area schools: Sidwell, Georgetown Day School, Maret (all in D.C.), and Burgundy Farm Country Day School (near Alexandria, Virginia). The friends said Sidwell was far and away the best and most prestigious, which is technically true. However, my parents knew *nada* about the subject, so they automatically wanted Sidwell whether or not it was the right school and best fit for me.

As I sat there in my patriotic thick cotton piqué mini-dress (white bodice, red empire waist, navy blue skirt), white knee socks, and orthopedic saddle shoes, engulfed in that big old dark maroon velvet sofa with a gray pillow on my lap, I couldn't have cared less about the school's putative cachet. Actually, I was hoping I wouldn't get in. For one thing, I'd heard the place was academically demanding, and I was really, really busy with other things; 1967 was an amazing year for pop music, and I wasn't sure how I'd squeeze that much "studying" into my already crammed music listening schedule, not to mention my magazine and book reading, not to mention all my TV show viewing. *Family*

Affair, Saturday Night at the Movies, and *Green Acres* were among my faves, though they often got interrupted with special reports about the six-day Arab-Israeli War (which my parents and I cheered when "we" won), Dr. King encouraging draft evasion from Vietnam, civil rights riots, and the world's first heart transplant. I was also trying to keep up with my weekly typed diary entries. I liked recording what was in my mind while listening to music sometimes; I had a little plug-in record player and a growing tower of 45s and LPs. That year I especially loved "To Sir with Love," "Ode to Billie Joe," "Somethin' Stupid," and "Girl, You'll Be a Woman Soon." Beyond my pop culture-dom, I was happy and doing well at Amidon with my *negritas,* with whom I'd walk to and from school. If I left Amidon, all that would cease. No more birthday parties with lemonade and coconut cake and potential black parents to adopt me, no more double Dutch.

Knowing I would be leaving, my fourth grade teacher, Mrs. McEachnie, inadvertently made it harder for me to say good-bye by being extra nice. As a farewell gesture, she asked if there was anything I'd like to do that we hadn't done. I didn't have to think. Papier-mâché. So the class spent a week making our fanciful *objets,* and it was as fun as I'd imagined, and not once did it ever remind me of six million dead Jews, thereby proving that doing crafts didn't necessarily lead to gas chambers and that Mami was nuts.

Why was my family like this? Was I adopted? That would explain a lot. But I knew I wasn't; my fingers and toes are exactly like Papi's, short and wide, not to mention our matching flan-size pores and acne-prone oily skin. ("Joo got dat from *hees* side, honey," Mami always told me.) Inspired by Hans Christian Andersen's "The Ugly Duckling" and William Butler Yeats's gorgeous and haunting poem "The Wild Swans at Coole"—one of the few grown-up Anglo poems I could sort of understand at that age—

I settled on creating a papier-mâché life-size swan. In my illustrated fairy-tale book the swan was white, and Yeats never
describes his "nine-and-fifty" swans by any color. So, being a
Jubana, which means you must have color at all times, I painted
my swan periwinkle blue—the only universally flattering color—
with a glossy black beak and big green eyes with a double row of
long black eyelashes (top and bottom) like mine. A gay friend of
mine in college who was obsessed with Elizabeth Taylor once told
me that Liz also has double rows—"upper *and* lower lashes," he
noted.

"Big deal," I said. "I have them, too."

He came in close and stared at my eyelashes, inspecting them
more carefully than he ever had before.

"You *do,*" he said. "Wow. Maybe it's a Jewish thing?"

"Cubans press fallen eyelashes together," I said. "You guys
break wishbones for luck. We put an eyelash on our fingertip and
press it against somebody else's. And whoever's got the eyelash
afterward has the luck."

Anyhow, my swan was a tour de force, if I do say so myself. In
fact, the boy who sat beside me, Maurice, was so impressed, he
gave me my first kiss on the lips, or at least that was his excuse.
But then Maurice tried to pick up my mysterious, beautiful swan
before it had finished drying, and I had to smack his arm, thereby
breaking the romantic spell, such as it was. Maurice got mad or
maybe it was just an accident, but as we were leaving class he
closed the door so hard on my hand that my right middle finger
broke. I got a splint and a cast with a shoulder strap and everything. Maurice couldn't even look at me, he felt so guilty.

My poor middle finger never did heal properly. To this day
there's a protruding bump on its left side. The finger incident
hurt, and it prevented me from writing for a long while, which was
the worst thing I could think of. God had given me God's vision for

my literary future and then some boy with no vision at all goes and breaks my Michelangelo Sistine Chapel ceiling finger. In the panel called "The Creation of Adam," Adam's inert finger awaits the spark from God's fingertip. How can you be a writer if you can't physically write? I thought of that years later when my beloved cousin Melanie became a hand model in Los Angeles. Whenever I'd fly out there, no matter what the occasion or weather, her stellar little $450-a-day hands would be encased in a pair of $500 quilted black leather Chanel gloves. I'd like to see Emily Dickinson in a pair of those. Anyway, I took the finger-breaking episode as a sign that Amidon should be a thing of the past. You can't stay in a place where boys break your bones.

I was hoping to go to Georgetown Day School. I'd had a good interview there, and the atmosphere was easygoing and nurturing. Maret was sort of in the middle. And Burgundy Farm, well, it was fine, but they had live chickens pecking and scratching around the grounds of the picnic tables where we'd eaten lunch. Mami instantly nixed payeengh top dollar for dat country cheet weeth dos *pollos*.

No *pollos* at Sidwell.

"It's fascinating you come from Cuber," Mr. Arnold said.

"From *where?*" I said.

"Cuber. Havaner, Cuber."

"You mean Havana, Cuba?" It was a rhetorical question. I'd heard poor dead President Juanito F. Kennedy pronounce the name of my birthplace the same way on TV during the Bay of Pigs and the Cuban Missile Crisis.

"Right, that's whut Ah said. Havaner, Cuber. Real neat! Spanish comes easily to you, does it?"

"Like English does to you," I said, my eyes on the clock on his desk. Like everything else in this office, the clock was wooden and plain. Spartan. Spartan is anti-Juban. It's Quaker. I stared at the slo-mo second hand. This was taking forever.

"So you're completely bilingual?" Mr. Arnold asked. "Perfectly fluent in both languages?"

"Perfectly perfect."

He went to the dark brown bookcase, pulled out a heavy hardback book, and laid it on my lap, opened to a map of the Caribbean. It was an eighth-grade-level geography textbook in Spanish.

"Recognize any of that?" he asked.

The guy may have been an Ivy Leaguer—a term I'd picked up back at the three other private schools I'd already visited—but I thought he was a complete idiot, or worse, disingenuous *because* he was a complete idiot.

"I recognize that like you would recognize a map of the United States," I said, hoping my smart-aleck tone would turn him off and jeopardize my chances of admittance. "You *would* recognize a map of the United States, wouldn't you?"

"Whyn't you go 'head and read some of that text to me."

Since Mami had taught me to read in Spanish from the age of two, this was a ludicrously simple test. I rat-tat-tattled away in rapid-fire *Español*. Just as I was getting to the tongue-rolling section on La Sierra Maestra, Dumbo said to stop.

"If you said *la maestra cierra*," I noted, "it would mean something totally different."

"Really?" Mr. Arnold said, scratching his head. "Whut?"

"*La maestra cierra* means the teacher closes. You know, like, a door. A book. A MIND."

"Okay," he said, taking the book. "We're in good shape. Thank you." His ears were so pink they were almost red.

"What did I just read?" I asked.

"How's that?"

"I can obviously read in Spanish but can you understand what I just read? How do you know I didn't make any of it up? I could've been reciting José Martí poems, for all you know. 'Los Zapatitos de Rosa,' for example."

"Huh. Well. Ahm listening for your verbal ease and acumen as well as your intellectual comprehension," he said, "not just literal word definition or foreign translation and interpretation."

Domestic translation and interpretation: *GET ME OUT OF HERE. NOW. Please, Yahweh, I'm begging you. In the name of Abraham, Moses, David, and all the major Old Testament Israelites, in the name of my Holocaust victim great-grandparents and all the women prisoners who stopped getting their periods from the sheer stresses, in the name of every Bay of Pigs mártire and all of Cuba's political prisoners, in the name of every descendant of every slave on every Southern peanut butter plantation, please Lord God, liberate me from this cahm!*

I stretched, yawned, looked out the window. On a big green field raced muscular, red-cheeked girls with messy ponytails, wearing gray sweatshirts, maroon kilts, maroon knee socks, and lunky-chunky athletic shoes. The girls were catching and chasing a ball with wooden sticks that had nets on top, like brawny butterfly catchers.

"Girls' lacrosse team," Mr. Arnold said, jutting his chin in the direction of the players. "What sports do you enjoy?"

"Where are black girls and boys?" I asked.

"Sorry?"

"Where are your black children students?"

"Oh, we've got a few."

"Not out there. I can see in my glasses. How about Latinos? Besides the janitors, I mean."

"Well, we pride ourselves on . . . So you play lacrosse at all?"

"What?"

"You like lacrosse?"

"I wouldn't know," I told him. "Never been there. Wisconsin, right?"

"Say whut?"

"OshKosh B'Gosh? Cheddar cheeses? Sharp, extra-sharp? Cows? Mami says the Midwest is a pit, like Puerto Rico. Actually, she thinks this whole country's *una mierda*."

"Well, hahaha. No. You see, here at Sidwell, we consider rigorous physical education as vital to our overall core academic curriculum as we do the Quaker ethos."

This was as indecipherable as that Mississippi psychiatrist Joe calling me a pistol. Did Mr. Arnold just say that you have to eat oatmeal to attend this school? Speaking of which, I was starving. Thirsty, too.

"Are we eating lunch soon? I can't think or talk anymore when I'm this hungry."

"Sure, we'll go grab a bite soon," Mr. Arnold said. "Ah think today's either mac 'n' cheese or Charlie's Special."

"Charlie's *what?*"

"Oh, it's special, all right. You'll like it if you like chicken. It's like a stew, named after one of our greatest coaches. You like chicken?"

"The dead or alive kind?"

"Dead or—?"

"I like *arroz con pollo* if psycho-maid makes it and not Mami," I said, meaning our maid, Rebeca. The first time I saw her was after they wrenched out my tonsils at Children's Hospital and on the way home Mami said we had this new maid who was making me something special. *Something special.* This could be good! I'd always had nice nannies. Maybe Rebeca, my namesake, will have made me cherry Jell-O and lemonade and bought some mint chocolate chip ice cream. That's all you can handle after a tonsil-lectomy. The pong of frying eggs and ripe plantains and rice hit me in the face as we walked through the door. Normally I'd have loved it; that's a typical Cuban lunch or light dinner. But not right after throat surgery. Mami introduced us and Rebeca stared me down with fierce black Andean Indian eyes. Ecuadorian eyes. She wore what appeared to be a nurse's white uniform minus the cap. She was a pygmy, barely taller than I. Her jet-black hair was wavy, shiny, and thick, sprouting out of her head like Greek myth coils. She had tucked a series of bobby pins in it on both sides to keep it back off her face. Her earlobes drooped from heavy dangling gold earrings. There was a golden crucifix on a chain around her neck. I got the picture immediately. Rebeca was one fearless wench.

The clock on Dumbo's desk struck noon. "Rebeca's an An-dean Eva Braun but she can cook," I told him. "Look, I really have to eat now. Please. Get me food. Or I might say ain't and my mother might faint and my father might fall in a bucket of paint."

He regarded me, stupefied.

"Double Dutch," I explained. "Not lacrosse."

I once asked my rabbi, Bruce Kahn, where God went during the Holocaust. Bruce said, "If we knew the answer to that, we'd be God. But there's more to it than that. I believe that God's power is all over, everywhere for us to take in and utilize. We blame God for our failures. God has given us the clearest possible understand-ing of what the sacredness of life is about and what we must do to safeguard it. And if we're told to stop brutality and hatred and murder, let alone genocide, and we ignore all the warnings we're receiving and all God's guidance to take action against it—and it happens, why is it that God is to blame? That is to say that human beings are less than puppets."

"Bruce," I said, "you are a god. Small G, don't get nervous."

"Nah. I'm a servant of God. But seriously, think about it: If you're given all the power, insight, and direction you need and you wind up with a tragedy on your hands, whose failure was it?"

"I don't know. Is this like that 'when bad things happen to good people' attitude?"

"Was the Holocaust God's failure or that of the human race? When you do everything you can to instruct a child to not do something that would hurt the child and the child does it anyway . . . And we're talking about children here, not adults."

"So we poor little Jew people are to blame?"

"No, I'm not talking about the victims of the camps but of the leadership of the world that had an awareness of the rise of Hitler and his genocidal message and who chose appeasement as opposed to prevention. So I think that God was rarely as disappointed in the human race as He or She was when the world wouldn't deal with the rise of Hitler and Nazis. I'm sure some Jews who died refused to face and deal with what was really happening—some led very narrow lives and were not that sophisticated. But the majority had no adequate awareness or options to do much different from what they did."

In my current Sidwellian context, that meant Mami and Papi chose to listen not to me but to their friends. And I had nothing to do but go there to that preppie cahm.

"The approach of the world is to not deal with the ethic that it espouses," Bruce added. "People will in lots of personal ways but when it comes to real decision-making, it's very difficult. There are usually other considerations than the morality of right and wrong: financial gain, political or some other kind of prestige. The expediency overrules those other ethics—justice, the saving of life, and imitating the goodness of God."

As usual, Bruce, a reservist naval chaplain, said a mouthful. I say that God and the parentals must have had something *extremely* mysterious in mind on the day Sidwell accepted me, one of the singularly worst days of my single-digit-aged life, and God knows I'd already had my share: Cuba and Cecilia and guerrillas and my stolen little red tricycle and Rebeca and my broken finger and tight shoes and black parents rejecting me as their own were all bad enough— and now *this?* Why didn't God go and smack Mami and Papi upside the head and make them send me to Maret or Georgetown Day School instead of to Sidwell? Manny Roman says I obsess over imponderables. I'm sure it's all the *café con leche.*

"I like Georgetown Day and Maret the best," I told Mami, checking myself out in the car's passenger vanity mirror. We'd just been to Robert England's hair salon on M Street in Georgetown, where, in a Twiggy-inspired impulse tinged with a soupçon of Vidal Sassoon sprinkled with a dash of Faye Dunaway's sleek Bonnie and Clyde bob, I'd had my waist-length hair lopped off into an adorable, boyish pixie cut like Twiggy's: short in the back and around the ears, parted on the side, with almost too long layers in the front that occasionally obscured my vision as well as my hideous glasses. I loved it.

"Eef Seedwells Frien' takes joo," Mami said, chewing her Juicy Fruit and exhaling Kool smoke through her nostrils like a dragonet, "joo go der. Ees de best. So. End of story."

To think that that cahm would be my scholastic pokey for the next eight soul-killing years . . . Isn't that the essence of tragedy, really, when you anticipate a disaster—and then it *happens?*

For all Cubans of my parents' generation, the collective Juban mind-set is Havana, circa 1955. And like children, certain exotic tribes, and mental patients, Jubans traffic in magical thinking. It's how they get through life. It's Blanche DuBois: "I don't want realism. I want magic!" Magical thinking, not to be confused with magical realism, permeates every aspect of living. In our case it has to do with nostalgia, fantasy, the belief that thinking equals doing and, above all, denial. As the Temptations sang, "I can make the season change / Just by waving my hand / I can change anything from old to new / Yet the thing I want to do the most / I'm unable to do."

Like make Fidel never have happened. That's impossible even for a mighty Temp' to pull off. Or even for the super-strong, super-fragile, and altogether extraordinary Annie Lennox, who covered that Temps song so gloriously on her *Medusa* CD, and who sings everything else so gloriously. How *do* black American soul singers and Scottish chanteuses and the other artists I love so much, those intimate strangers both dead and living in whose beautiful recorded, written, painted, cinematic, psychiatric, culinary, and even sartorial voices I can privately and profoundly lose myself and find and feel my own joy, grief, solace, beauty, despair, desire—all the otherwise inexpressible human emotions—be so different from me and move me so much that I feel us connect? How is it that Shakespeare understands *me*? Or Manet? Or Manny? I could go on and on about people who make my life worth living. (As my writer friend Mr. Lewis Lawson says, "Some of us writers are taker-outers and others of us are putter-inners." I'm obviously in the latter camp.) But here's the question: Are Jubanas really so different from everybody else?

And here's the answer: a little bit, yeah. The parentals' child-like clinging to the magic of denial so infused the atmosphere as I grew up that I didn't realize I was inhaling it like secondhand Kool smoke. My exiled Cuban family's own little Motown-inspired mantra goes like this: *It's not real if I say it's not. If I don't look at it or think about it, if I don't accept it, I can will it and wish it away. Any bad thing will be as though it never existed and only everything lovely and pleasurable will go on forever without end. If I believe in it, every day and every moment can be the Fourth of July, only better and grander, with much bigger and more dazzling fireworks. Nothing comes at any real cost because I'm entitled to it because of who I am.*

Conveniently, denial is a portable philosophy. Pack it and go. Since Castro allowed us *gusano* traitors only a few suitcases apiece, the good news is that denial hardly takes up any room. If you want to really protect it, you can do what Barry Fletcher did with his prized blow-dryer when the D.C. hairdressing champ— whom I wrote a story about for the Style section of the *Washington Post* in 1994—headed off to international hairstyling competitions: He wrapped it in a towel and *then* packed it in his shoulder bag.

Mami also wrapped towels around her prized possession— denial, never *self*-denial—but hers were Towels, not towels. My parents had received, among many other beautiful things that Castro's gangster guerrillas eventually stole, a huge set of monogrammed bath towels. They were solid and striped, in every color, hand-loomed of the finest pima cotton. Today they're in tatters, thin as onion paper. Still usable, but rarely in use. Mami's bought many other towels, bigger ones and maybe even better ones, in different American stores throughout the years. Mostly those are used because, over time, my parents' wedding towels have aged into fragility. They lie in wait, folded neatly in thirds in perfect vertical stacks in the linen closet.

A fragrant mausoleum of emaciated Cuban wedding towels. A Juban towel museum. Tomb of the unknown towel.

One time, I suggested to Mami that she let them *go*.

"Let dem GO?" she shrieked. "Let dem go *where?*"

"Out 'DER,' " I said. "Isn't that what you always tell me *I* should do? 'Get out DER'?"

She flashed me The Look, blew Kool smoke in my face, and slammed a rather attractive stolen London restaurant ashtray down on the kitchen counter, cracking the impacted topaz-colored imported tile. (Cubans are tile-obsessed. It's a Spanish—as in, *from Spain*—décor motif.)

"Dos towels, dos towels are EET. Okay? Deyr great! Deyr de best ones! Deyr my whole damn life before dat fohkeengh Castro. I'll *never* get reed of dem, *never,* so *fohk heem.* Motherfohker. An' joo can forget eet."

So glad we straightened that out. Thank God we're not bitter or anything. God forbid anything petty should interfere with our assimilation.

In fifth grade the only control I had was over my body. So I asserted myself with my groovy haircut, just like a good Hemingway heroine. (In Hemingway you can always tell there's a sea change in a female character when she cuts her hair.) Mami had initially refused to pay, insisting I was "maykeengh de meestayk of a life-time." In other words, straight men prefer long hair. In other words, an up-do is more elegant than a boyish pixie cut for a *tafetán color champán* bride. In other words, just like taking ballet and drama, me having long hair was vicarious for Mami, who has almost always worn her hair short. She's never had any patience for anything, and with her hair's nice color but wispy texture, she'd have had to really work it to make it look halfway decent if it were long.

"*¿Si a ella le encanta tanto el pelo largo,*" Rebeca said, "*pues entonces por que ella misma no se lo crece?*" If she loves long hair so much, then why doesn't she grow it herself?

Good point. Did I detect an edge in her tone? Was this pygmy watering the seed of my budding ambivalence toward Mami? And if so, to what end?

"Ella no lo crece porque no le quedaría bien," I told her. She doesn't grow it because it wouldn't look good on her.

Rebeca shot me a significant look, the silver blade of her knife shining. If she could have raised one eyebrow, she would have. We were in the kitchen, Rebeca slicing peeled, seeded cucumbers and me sitting on the high stool and eating the discarded wet strips of seeds sprinkled with a little salt, one of my favorite treats.

"¿Qué tu estás diciendo?" I asked. *"¿Que ella QUIERE que yo luzca mal?"* What are you saying? That she WANTS me to look bad?

Rebeca smiled knowingly and whacked a naked cucumber with appalling gusto. See, this was the thing about her: I still couldn't stand her—she was a misogynistic control-freak Jesus-freak radical traditional Hispanic female with all of the retro bull-shit and none of the assimilation or wit. But! She did have some good insights from time to time, plus she kept me, Señorita Lonely, company. My parents had each other, and my brother Eric would soon have a new redheaded baby brother named Big Red Al to bond with. Whom did I have just for me? Since Cecilia was gone, no one. So Rebeca kind of won by default. This latest insight of hers about my mother and me and our hair was . . . what, exactly? I wasn't sure, and Rebeca was always shrewd enough to pique your interest and then go forth and say no more, a savvy provocateuse who withdraws just at the moment of climax. She must've been the life of the village back in Quito. Well. Rebeca may indeed have come from penury and she was uneducated but she was no *bruta;* she was a hard-core survivor who knew which side her Wahndehr Brayt was buttered on. So she'd tease me with an intriguing morsel about Mami Dearest, who paid her her salary—and then go whack a cucumber. It was perfect in its Freudian simplicity.

And my new hair was perfect in its precise geometric simplicity. Up until Twiggy my style icon had been Nancy Sinatra, especially the way Frank's firstborn looked on the cover of her 1966 *Boots* LP. The black-and-white striped bodysuit, red leather miniskirt, and coordinating red leather go-go boots with the roll-down tops—so great. I also loved her *How Does That Grab You?* LP cover look—oversize camel fisherman sweater to just barely below the butt, possibly tights (it's hard to tell for sure), and brown leather knee-high boots. My other fave was her *Nancy in London* look, which even now looks au courant: brown newsboy cap, straight highlighted blond hair, chunky red turtleneck, skinny dark blue jeans, suede camel boots. Yep, Nancy definitely had it going on as far as I and my fashion and hair choices were concerned, until Twiggy, with her velvet painting Third World starving orphan hyperbolic eyes and fake eyelashes, caught my eye. But then I wanted to be Faye Dunaway's Bonnie Parker and I grew my hair back out into a pageboy so I could wear a white or black wool beret (tilted at a rakish angle, of course) and lie on my bed (the wrong way, with your cutlets on the pillows and your head where your cutlets go) and bang on the footboard with my tiny fists in frustration over the confines of my real life. But then Mia Farrow looked so elegant and chic in *Rosemary's Baby*, even when she sweated in un-air-conditioned Manhattan phone booths, that I cut it all off again. Then I let it grow out after seeing the lit-from-within Olivia Hussey in Franco Zeffirelli's *Romeo and Juliet* and I kept it growing for two years, through Ali MacGraw's foul-mouthed yet perfectly dressed Radcliffe Catholic beauty in *Love Story* in 1970. And no matter what my hair situation was, there was always Agua de Violetas in it.

One day while I was Mia-fixated, I bought a magazine with her angelic face on the front. The feature story was about how Frank Sinatra had abruptly divorced the refined waif for having shorn her locks, allegedly yelling, "I married a girl and woke up with a boy!" It was raining lightly out and I was standing in front of the supermarket with our bagged groceries, waiting for Mami to pull up the car. I was completely engrossed in Mia's marital and beauty travails, thinking how Frank could be a Latino with that Neanderthal short-hair-makes-you-a-dyke attitude. I heard a loud quintuple honk, forgot there was an elevated curb, approached the car—I was still reading the article—and promptly fell on the slippery pavement. Poor Mia's face was a dirty mess. I immediately wiped it off with my arm, over and over.

"What are joo DOOEENGH down der?" Mami screamed. "Joor knee ees bleedeengh to de deaths an' joor cleaneengh Mia? ¿Tu 'tas loca? ¿Qué RAYO 'tas haciendo?" Are you crazy? What the HELL are you doing?

"I'm prioritizing!" I cried. "That's what Americans DO!"

Mami would never have screamed like that had I been holding an *Hola!* instead of a *Rona Barrett's Hollywood. Hola!* is Mami's and our family priest Máximo's favorite Spanish-language glossy. The oversize magazine obsesses less over Tinsel Town and more over Actual Royalty of the European kind. Had, for example, La Princesa Carolina de Monaco's extraordinarily beautiful face been soiled in front of Giant Foods instead of Frank Sinatra's soon-to-be ex, Mami would have been the first to wipe off now-poor-dead Grace Kelly's daughter *tout de suite.* Unlike me, Mami was never influenced much by American pop culture or the evolution of American fashion and style. She's had the same three-feet-long square-tipped fingernails painted bloodred or opaque white since she was a girl in Cuba (must have been the height—or depth—of chic in 1955 Havana), and nothing will ever make her change that. If you tell her that that look is beyond disco passé, her stock replies are, "What ees joor poin'?" or "I don' geev a fohk. I lohvee."

I, however, *deed* geev a fohk. At four I'd been seduced forever by *Seventeen.* And now at nine years old I'd long outgrown my *Archie* comic books and had moved on to fashion and celebrity 'zines. So much so that I'd actually begun purchasing annual subscriptions with my allowance because it was cheaper than buying them all full-price at the store. These days I'm down to a mere twenty-four subscriptions; it used to be twenty-seven. *Mademoiselle*—sob!—folded. I didn't renew *Condé Nast Traveler*— I only leave my apartment by force—and *Rolling Stone* hasn't been my thing since I began shaving my armpits and legs. When I see all my new magazines in their plastic covers, heavy and rolled up and layered and smushed inside my mailbox, I feel euphoric. Little did I know back then that this fascination would be the perfect preparation for becoming a features reporter at the *Washington Post* and later, a freelance magazine writer. The fiancé and I had this joke: I'd be telling him something about *Sex and the City* or Narciso Rodriguez or Sephora or glittery lip gloss or the latest hair removal or defrizzing techniques or chandelier earrings or anything else of crucial import that I'd 'zine-gleaned, and he'd go, "Whuuut?" And then we'd both say, "Hard news, Dinosaurio," because Paul'd spent his entire professional life as a hard news reporter and editor—the newsroom chasm between hard news and features people is cosmic—and because my nickname for him is Dino(saur) or El Dinosaurio or E.D. for short. Paul's Mesozoically big and basic and, as Mami puts it, "sohleed." There was an obscure girls' song in the late seventies whose refrain was, "I like 'em big and stu-pid, I like 'em big and REAL dumb." Works for me. As Chanukah stocking stuffers, I once got Paul two refrigerator magnets. One says "I ♥ My Penis" and another one has a picture of a huge sliced bologna that says "You're not too smart. I like that in a Man." Paul cherishes them. He's so smart. He's so stu-pid.

To my complete confusion, my impudence during the Sidwell Frenzy interview backfired and I was accepted. Maybe WASPs have an ironic sense of humor? Maybe they view impertinence as gumption? What did I miss? I quickly discovered I was profoundly ill-equipped and tragically unprepared to navigate this bizarre late sixties sea of a Quaker private school, in the midst of Vietnam and with Watergate soon to come. Nothing computed: Turtlenecks had to be from Talbots and have embroidered whales on them; espadrilles came from Pappagallo in Georgetown; Ivy-grad teachers who looked like robust, predissolute Ernest Hemingways and Martha Gellhorns taught Ivy-bound kids who looked like Ralph Lauren models of Ivy-grad parents who looked like Donald Rumsfeld, H. R. Haldeman, and Jeane Kirkpatrick (all of whom actually did have kids there), and everybody would look like John Cheever and Lilly Pulitzer when they were old; if you lived in D.C. you lived in Georgetown, Woodley Park, Spring Valley, Foxhall, Friendship Heights, Wesley Heights, Palisades, Glover Park, Tenleytown, or Cleveland Park; in Maryland, you lived in Kenwood, Bethesda, Chevy Chase, or Potomac; if your parents were alcoholic anti-Semitic Republicans, you lived in McLean, Virginia; you were into lacrosse, archery, tennis, soccer, basketball, cross country, crew, squash, football, softball, track and field, swimming and diving, field hockey, and wrestling; you had beautiful hair, perfect skin, a trust fund, and serious drug and alcohol issues.

By contrast, I felt like Edvard Munch's screamer.

Then I thought of Chet Baker. Beautiful, brilliant, doomed, drug-addicted Chet Baker. Except for the drugs, I could relate. I specifically thought of the way the jazz trumpeter played and sang "Let's Get Lost": "Let's get lost, let them send out alarms . . . /Let's get crossed off everybody's list . . . /And though they'll think us rather rude, /We'll tell the world we're in that crazy mood."

I'd heard it for the first time at Rhoda Simpson's house. She was one of my Amidon classmates whose parents I'd tried in vain to seduce into adopting me during a birthday party. The Simpsons said Chet Baker was so good he could be a Negro like them. Except for Frank Sinatra and Peggy Lee, Chet Baker was the only cracker in their entire vast music collection. ("Let's Get Lost" was a top contender for the fiancé's and my First Dance as husband and wife. The other was Etta James's "At Last." Both are wonderful; the latter is a slower song and therefore easier for E.D. to dance to, what with his big paws or toes or whatever it is that dinosaurs plod on.)

"Gigi's goheengh to Seedwells Frien'!" Mami announced.

It was a Sunday in late summer. Dozens of neighbors and other friends had come over to celebrate her and Papi's August birthdays. Mami prepared her signature Sunday afternoon meal: *huevos a la Malagueña,* baked Eggs Málaga. It's a traditional Spanish casserole of whole eggs baked in a bed of *sofrito*— chopped onion, bell pepper, and garlic sautéed till tender in olive oil, the holy foundation of all savory Cuban cooking—mixed with sherry, tomatoes, and pimientos. You sprinkle early sweet peas on top (Cubans only use LeSueur) and lay on canned asparagus tips (Cubans prefer canned, according to Mami, because when canned produce was first introduced in Cuba, it was considered technologically advanced and more sophisticated than the raw stuff). Now the classic way to make this dish, which I've heard is actually quite tasty if you bother making it correctly (which, God forbid, requires more dan three eh-stehps), is to drizzle each egg with melted butter, fortify the rest of the casserole with chopped ham and boiled shrimp, and then bake it *lightly,* just until the egg whites are set but the yolks are still soft.

But no.

It's not that Mami excluded the pork and shellfish because they're *trayf.* (Rebeca fried us bacon every day of our lives, Baba Dora's specialty was ham sandwiches, and Mami took me to go sit on Santa Claus's lap every Christmas. Until I got my period at age twelve, that is, after which I had to stop going because continuing to sit on Santa's lap would be, as Mami said, "eefee." Not because I was physically blossoming and turning voluptuous. Not because Cuban parents, like all Latin parents, are so conservative and overprotective of their children, especially their vulnerable, virginal daughters. Not because—here's a rad' concept—we're a *Jewish* family who respects the *goyish* tradition of but does not actively engage with Santa. Not because the shmuck at the mall wearing a rented Santa costume might actually be an alcoholic kiddie pre-vert with a rap sheet longer than his gift list. No. It's because, "Joo might eh-stain Santy Clohs—joo bleed like krehsee, wheech I can never understan' because I never had dat, but joo are so exaggerated een everytheengh—an' den dat would be horeebl! Because eef Santy ees not wearing rayt pants, less say hees wearing, like, white pants or sometheengh like dat, den eet would be, like, a total contrast! Das a no-no.")

So.

What it is, is that Mami would have had to go out to an actual *grocery store* and actually *purchase* ham and shrimp, which are not only more expensive than a carton of Kool kings but require actual *preparation.*

It's not that the *huevos a la Malagueña* got no butter on their respective tops because we ran out of butter or we don't believe in butter. It's that the butter has to be *melted*, and why would you stand there melting butter and then have to deal with yet another item to wash when you can simply . . . not?

It's not that Mami overcooked the dish to the point of *huevo* fossilization because she liked her *huevos* hard. It's that when you insist on having a TV set in every room of your house (why else go on living?), you're bound to WATCH TV AND NOT THE OVEN.

Guests are always too polite or too ignorant or too hungry to say anything negative about Mami's *huevos a la Malagueña*. Papi had bought several dozen assorted bagels and a nineteen-pound slab of cream cheese. Mami made tuna salad and egg salad (both actually fit for human consumption) and there was a huge hunk o' Jarlsburg cheese and a bowl of green grapes. There were cheese blintzes and *empanadas de picadillo* (turnovers stuffed with ground meat). Papi made a pitcher of sangria and there were the usual twenty-nine pots of Cuban coffee with the requisite three thousand pounds of sugar mixed in.

I grabbed a TaB out of the refrigerator, spread some cream cheese on half a bagel, and sat down with the company as Eric made the shoe shine rounds in his squeaky Pampers.

"So, Gigilah," Neil Greene said, "you must be happy about your new school." He was a good-looking architect who lived a few doors from our town house

"My new school?" I said, stalling. I slowed down my chewing and pretended the thickly spread cream cheese was sticking to the roof of my mouth, making it impossible to talk.

"Sidwell Friends," Neil said. "That's a big accomplishment, bubbeleh. How many new kids do they accept every year? What is it, David, something like seventeen percent of all applicants? Eleven?"

"I really don't know," Papi answered, refilling Neil's sangria glass. "But for the tuition those ganefs charge, it had better be . . ."

"Seedwells ees de BEST," Mami snapped. "Gigi ees threel-ed to go der. THREEL-ED. Right, *mumita?*" She cupped my bagel and cream cheese–stuffed cheek in her cool hand. "Nex' weekend we're goheengh choppeengh for new fall outfeets. Ees so much fun, ees eencrehdeebl!"

"I have to go now," I said, stepping over Eric, who was on his hands and knees, digging in the brick-colored shag carpeting for an errant quarter. I ran upstairs to the aqua sanctuary of my bedroom. I was breathless and a little dizzy. My heart was racing. I had a terrible presentiment about Seedwells. But the fact of me going there was inexorable. I lay on my bed and curled up into myself like a fetus, hugging my second pillow (Mami's always believed in more than nineteen of everything, hence many pillows on every bed at all times). I wished I still had my *gindaleja*, I really needed it now. But to commemorate my fourth or fifth birthday— God, what was I *thinking?*—I announced to Mami that *gindalejas* are kid stuff and I was a big girl now. She asked me if I was absolutely sure and I said yes and we threw it away. In the middle of the night I ran to my parents' bed and jumped on Mami, screaming for her to get it back.

"What?" she said. "De *gindaleja?* Das gone, honey. I ask-ed joo are joo choor an' joo said jes."

"I changed my mind!" I cried.

"Sorry, *mamita*. Ees a goner. Go back to joor bayt, okay? Jool be fine. Nighty-nighty."

"Oh my God. I want it back!"

"Nighty-nighty!"

"It was a moment of weakness!"

"Das great, honey. Nighty—"

"Well, can you at least make me some *leche evaporada caliente con azúcar* [warm evaporated milk with sugar]?"

"Dahveed," Mami said, poking Papi's shoulder. "Wake OHP. *La niña quiere leche. ¡Despiértate, coño!*" The girl wants milk. Wake up, dammit!

Papi dutifully arose, and while he heated a can of evaporated milk, poured in the requisite fifteen cups of sugar, and stirred, I inspected all the garbage cans in the apartment—we had many; Jubans are incredibly trashy—to make sure Mami wasn't lying. She wasn't. *Coño!* I told Papi that I was fully prepared to slide down the garbage chute down the hall to exhume the *gindaleja* from the Dumpster. Papi said, *"Tu 'tas loca? Esa cosa 'sta llena de microbios. Olvídate."* Are you crazy? That thing's full of germs. Forget it.

Like Mami, he always was kind of literal.

We walked back to my bed, and Papi sat on the edge as I drank, his eyelids drooping. *"¿Ya? Acába allí, gorda, po'que 'stoy muerto."* All right? Finish up over there, fat girl, 'cause I'm dead.

Even tired in the lamplight in the middle of the night, my father looked handsome, like a deposed Latin American head of state with soft, kind, warm, sweet espresso-brown eyes. The fiancé called Papi *el caudillo,* which means "the leader," although it was more commonly used as a title for General Franco. (Paul called Mami *la esposa,* which means "the wife.") I gave Papi the empty glass and kissed his cheek. Then I started crying all over again, thinking about the lost *gindaleja.*

"No llores, bobita," Papi said, holding two tissues up to my nose so I could blow. *"Más se perdió en Cuba."* Don't cry, silly girl. More was lost in Cuba.

It all seemed like a long time ago now. But the memory was comforting, that feeling of warmth and fullness and *bienestar,* well-being, that the sweet milk and my father watching over me with his ever present neatly folded tissues gave. *The memory of all that, no, no, they can't take that away from me.*

I pushed away my pillow, rolled over on my stomach, and tapped on the window, daring to disturb the wasp universe. Right on cue, the winged *sohkehrz* buzzed and swarmed, flinging themselves with malicious *ping-ping-pings* on the pane. They repulsed me. Tomorrow I'd sic Rebeca—cheaper than Orkin and twice as sadistic—on 'em with a ladder, a hose, bug spray, plastic gloves, and the Ecuadorian broomstick she'd flown in on. That'd show those ugly Amehreecahn wasps who was really boss. In the meantime I'd recall *Shall We Dance.* I had watched the movie on a rainy Sunday afternoon on TV with Mami while Fred and Ginger danced to Ira and George's enchanting words and music and I handed Mami her manicure accoutrements one by one like a veteran operating room nurse. Mami had accumulated so many that containing them all required three gigantic picnic baskets. First we prepared: Ashtray. Cigarette. Lighter. Espresso. Then the Main Event. Acetone. Cotton balls. Emery board. Cuticle remover. Orange stick. Base coat. Color. Color again. Top coat. Once her nails were done I'd light Mami's cigarette for her while it was in her mouth so the lighter wouldn't mess up her shiny wet nails.

"Joo know de Gershweens were two Jews, right?" Mami said, accelerating the drying process by exhaling Kool smoke on her nails. They were ice white. Snow white. Albino white.

"Mm-hm," I said distractedly. I was engrossed in a new *Vogue* with Lauren Hutton on the cover. Tio Bernardo's second-born, a drop-dead gorgeous daughter named Lishka (he "created" her name), looked a lot like her. Dadeland Miami male drivers would crash into Burdines if Lishka was walking down the street. For a Jubana she was unusually tall and willowy, always tanned, with wavy-not-frizzy naturally highlighted dirty blond hair, huge hazel eyes, flawless skin, and a body worth giving up *puerco* for. Plus

Lishka was recognizably human, personality-wise. I loved her.
She was always clashing with Tío Nano, which I really respected.

"Doesn't Lauren look like Lishka?" I said, showing Mami the
photograph.

"Please. La-La [Lishka's nickname] ees so much preetee-er
dan her. An' La-La doesn't have dat gap een her fron' teeth. Oh, I
just looohv my nails! I looohv dooheengh dem. Ees my tehrapy."

"For what?"

"Life," Mami replied.

"Therapy like my *gindaleja?*"

"Jes."

"So now that mine's gone, what do I take? Do I need therapy?"

"Joo could smoke or sometheengh like dat."

"Maybe writing can be my therapy."

"Right but smokeengh ees a lot easier an' much more fun.
When people say, 'Why don' joo queet?' I go, 'I would not do eet eef
I deed not lohvee. Thank joo an' fohk joo.' For now johs keep on
dreenkeengh de Knox, okay? Joo are, right? Because joo can
transcen' all de coils of de mortahleetees an' de deesahpointmen's
by lookeengh goo'. Every *refugiado cubano* needs sometheengh.
Fohkeengh—"

"I know. Fucking Fidel Castro. Hitler's demon spawn."

"Joo got eet, honey. Joo can blame eet all on heem an' hees
dehveel father, Adolfo. Because de facts are de facts. Includeengh
de need for tehrapy an' long, hard nails."

Poor Mrs. Dorothy Blanchard (A.B., University of Nebraska; Co-
lumbia University). She just didn't know what to do with me or
her new scarf. Mami bought my Sidwell Friends fifth grade
teacher a faultless silk scarf in a primary-colored Frank Lloyd
Wright design from Lord & Taylor. It was a getting-to-know-you

gesture, a more sophisticated apple for the teacher. It was also a Juban bribe. Translation: Joo can be bought. Now let my baby slide.

Implicit Mrs. Dorothy Blanchard response: No, I actually can't. No, I actually won't. Where are you people FROM?

My perplexing predicament: Shit. The old silk scarf trick may not work here. Mrs. Blanchard looks like Pat Nixon, but with less warmth, spontaneity, wit, and enthusiasm. She expects me to know how to do fractions. I just recently mastered papier-mâché, for God's sake. I was doing swans. Whenever I attempt fractions my head gets hot like an overheated car engine and I have to pull over, cool off, and refuel with a nice *café con leche* and a lightly toasted strawberry Pop-Tart. These Gwyneths and John-Johns don't seem to get overheated like I do. There must be a paper somewhere with the rules on it, a manual that would tell me step-by-step how this game is played, the same way I learned which nail tools to give Mami in which order. Can academic achievement be like a manicure? In steps? Nobody tells me anything except when I do something wrong. Mrs. Blanchard just keeps giving me demerits for talking too much. AFTER MAMI GAVE HER A FUCKING FRANK LLOYD WRIGHT SILK SCARF. Really, really tacky! What's more, that scarf didn't buy me *anything.* Who's going to help me now? Certainly not my refugee parents. They're as lost as I am. They can't even help me with my fractions homework. They don't know what it is. Mami says she never heard of it.

If you go to the Sidwell Friends School Web site today, it will still be, exactly thirty years after my high school graduation, tragically all maroon and gray. But you will also find something even more shocking and telling about how it's entered the twenty-first century than the upgraded tuition ($19,975 per year for lower school and $20,975 each for middle and upper school): the Hola Corner. I love and covet the Hola Corner. It's the lower school's three-year-old mandatory *Español* program for all 290 of its adorable, privileged, pre-K through fourth grade *estudiantes*. You can even watch video clips of the tiny tots singing songs *en Español*, tots who will eventually go to Harvard and run this country bilingually. As a gay Nuyorican friend of mine says, "It's the new Spic Chic, darling. We're finally hot!"

Maybe we are. But Hispanics and Hispanic style were anything but chic and hot to most North Americans in 1967, and the North Americans running Sidwell Friends were no exception. Frustrated all fall by Mrs. Blanchard's lack of puppy love, Mami suggested I invite *la vieja*, the old woman, over for a Cuban winter dinner, a final, desperate Juban attempt at warming up and seducing *la gringa* with marinated pork products. After all, all my previous teachers had been over to the house. I really couldn't imagine having (or wanting to have) Pat Nixon there, though, chowing down on *puerco asado* and *frijoles negros* and *plátanos maduros*, foods she'd no doubt consider Fifth World sordid. So I sidestepped a direct invite by asking Mr. Arnold if that's "done" at Sidwell.

"Well," he said, Dumbo ears reddening, "it doesn't hurt to ask. But remember, if you invite her you'll have to invite her spouse, too."

That night at dinner I told Mami, "If you invite my teacher you have to invite her dog."

"Her dog?" Mami said. "Das really weird. I mean, lohv doggies, but steel, das really weird."

"I know but that's what the principal said. It's a kind of dog called 'spouse.'"

"'Spouse'? I never heard of eet. Must be, like, an Amehreecan breed das not dat goo'. *¡Ay! ¿Papito, que tu 'stas haciendo?"* Oh! Little Daddy, what are you doing? Underneath the dining room table Eric was manically polishing our shoes—for money. He'd gotten so carried away that he'd smeared the top of Mami's narrow white foot with thick black shoe polish.

"That child needs help," I said, curling my legs up under my butt to avoid the same fate. I sipped my TaB and picked stray pieces of *picadillo* and grains of *arroz blanco* off Big Red Al's hair, cheeks, and high chair tray. *Picadillo,* traditionally served over white rice with a side of fried ripe plantains, is a savory sautéed dish of ground beef, crushed tomatoes, sherry, cubed potatoes, dark raisins, and olives. It was one of Rebeca's many traditional Cuban specialties. Big Red Al had moved on to moisturizing his face and décolletage with a very ripe, very soft, sweet, and squishy fried plantain.

"I theenk dees keedees should go to Seedwells, too," Mami said, referring to my brothers. *"Qué te parece,* Dahveed?" What do you think, David?

"We'll see," Papi said. "Maybe."

I could barely contain my snicker.

"What?" Mami said. "Why ees dat fohnny?"

I had to think fast. In drama it's called *improvisation.* Following a successful run at Stage Studio as Helen Keller, we had moved on to Carson McCullers's *Member of the Wedding,* in which I played Frankie Adams, a lonely, alienated Southern twelve-year-old who alternately clings to and rejects her black maid. Perfect casting, *n'est-ce pas?* Let's face it, the part was not exactly a stretch for me.

"The fact that I go to Sidwell means Eric and Big Red Al will never get in," I said, trying to sound witty and ironic and self-deprecating. Hey, it wasn't a total lie. I *did* make Sidwell second-guess their decision to accept me on a daily basis.

Mami flashed me The Look.

"Wanna hear what I learned in school today?" I asked, anxious to change the subject and her terrifying expression. I cleared my throat and sipped some TaB. "Ready? Hell-ooo?"

Mami reached for an ashtray. It was a beauty, too: forest green with gold leaf and a lion's head insignia. Mami had snatched it and some lovely silverware from the bar at the Shoreham Hotel on Calvert Street in Northwest D.C. She and Papi went there a lot for cocktails and late suppers with Valerie and Walter, who were best friends with the swank hotel's owner Bernie Bralove, whose father Harry had built the place in 1929. Valerie and Walter lived in the opulent Rock Creek Park apartment building right by the Shoreham, and the Braloves lived nearby in a huge, gorgeous house. (I'd never been there but Valerie had told me it was really nice.) Bernie's second and current wife was Alice, a nice shiksa who taught ballet at the Washington School of Ballet just down the street from Sidwell's upper school campus. I was still taking ballet in über-gauche Southwest and had made my stage debut the previous December. I was a dancing snowflake in *The Nutcracker*'s snow scene and a marzipan in the second act, with a beautiful pink tutu all festooned with tiny flowers. Considering that I wasn't allowed to wear my hideous glasses onstage, I thought I did pretty well, though I occasionally collided with Lisa La Bicha, who was a lowly peppermint candy cane.

" 'O come let us adore Him,' " I sang to my bewildered family over the *picadillo* pieces, " 'O come let us adore Him, o come let us adore Him, Chri-ist the Lord.' "

"Wh-what?" Mami said, choking on her Kool smoke and coughing. She slapped her chest with her outspread palm as her eyes watered. "Wh-who? Wh-who are we a-a-a-doreengh?"

"Christ. *Chri-ist the Lord.*"

"*¿Tu 'tas loca?*" Mami said, slowly recovering from the Kool attack and dabbing her watery sunflower eyes with Papi's proffered folded tissues. "We don' adore Christ! Hees not our Lor'! De only Lor' right now for joo an' for me ees de one up der een de blue eh-skies an' de one down here weeth de an'-Taylor!"

"Lord and Taylor are Jewish?" I asked, confused. "Which one, the Lord or the Taylor? Or are they both?"

"Dahveed?" Mami said beseechingly to Papi.

Papi shrugged, sipped his espresso, and contentedly ahhhed, his eyes far off and away like one of those *New Yorker* maps of Cuba as the center of the universe just beyond the frontier of the Potomac River and our front door.

"Look, joor Lor' ees our Lor'," Mami continued, "an' our Lor' ees de Jeweesh one!"

"So the Taylor isn't Jewish? But the Christ was. Definitely."

"Dat ees SO not de poin'," Mami said. "De poin' EES dat we are payheengh all dees beeg bucks to dat damn school an' den joo come home seengeengh abou' Jesúcristo. *¿Tu 'tas loca?*"

"Eric and Big Red Al will sing for Jesúcristo too if you send them there," I said. "Good luck to you."

"Dahveed!" Mami cried. *"¡Has algo!"* Do something!

"¿Qué coño tu quieres que yo haga?" Papi said. *"Yo ni coño sé de que coño tu 'stás hablando. Lord y Taylor, Jesús Cristo—¿qué coño? Estoy, pero, perdido total."* What the hell do you want me to do? I don't even fucking know what the fuck you're talking about. Lord and Taylor, Jesus and Christ—what the hell? I'm, like, totally lost.

"Joo know what?" Mami said. "Ees time to do some beeg-time seenahgogh choppeengh. Like, NOW." We loosely belonged to a Reform temple near our Southwest town house, but it was tiny and didn't offer much in the way of Sunday or Hebrew school.

"Yo, bitch," I said to Mami. "It wasn't MY idea." As in, to send me to this charming Quaker school where Juban kiddies are forced to learn classic Yuletide ditties in order to sing them to residents of the Home for the Incurables, a Boo Radley sanatorium just up the street from the middle school on the corner of Thirty-seventh and Upton Streets, Northwest. At Christmastime the nurses would wheel out the superannuated vegetative patients, whose laps were covered in heavy crocheted afghans. Once they were all lined up on the balcony, we'd sing carols up to them. It's the Quaker Way: community service, civic participation, peace, etc. I once asked my teacher if we could try the Chanukah song about the dreidel or maybe fry up some nice greasy potato latkes with the applesauce and sour cream and bring it over to the incurables, maybe share a smoke with them as Mami did with the St. Elizabeths patients. The teacher regarded me like I was the Juban *anticristo.* My classmates convulsed in hysterics.

The Frenzy kids mistakenly thought that my suggesting a more inclusive holiday experience meant I was making fun of the incurable patients, patients they themselves were mean about, afraid of, and creeped out by. Patients they themselves feared they might someday become. *But youth is cruel and has no remorse and smiles at situations which it cannot see.*

Mrs. Blanchard and her spouse dog awkwardly declined the invitation. For whatever reason, the Jubanese Way just wasn't cutting it. So I opted for rebellion against the status quo. There were a few choice windmills down those elite Anglo hallways I could expertly tilt at. If I was gonna be shipwrecked on this preppie sea of embroidered whales, if I was going to fail—and this was a given—then I'd have to be the best failure these Quaker crackers had ever seen. Bangs, not whimpers.

Years later, too late to be of any use for Ivy League college admission purposes, one tweedy pre-Gramps Jewish-American Ivy

League shrink named Dr. Band—whom I nicknamed Dr. Band-Aid—posed the obvious question: "So one could say you were an achieving underachiever during your time at Sidwell."

"One could," I agreed.

"With no direction," he added.

" 'With no direction hooo-me,' " I sang. " 'Like a rolling stooo-ne.' "

"What's that?"

" 'You've gone to the finest school all right, Miss Lonely, but you know you only used to get juiced in iii-t . . .' "

"Come again?"

If you have to explain Bob Dylan lyrics to a person, there's really no point.

"I do have a plan," I said. "My plan is to have no plan. I expect to be dead before I hit thirty. That's about as planny as I can get."

"You *hope* to die so young? You *expect* to?"

"No," I said. "I just think I will. I'm a fatalist. What if just having fabulous nails and smoking isn't enough of a reason to live? Then what do you do? Even if Castro dies tomorrow—and he won't—the damage has been done, you know? Sorry to whine, but it's inexplicable to a non-Cuban. What are we gonna do when Castro ever dies, just pack up our monogrammed bath towels and pick up where we left off? I think Cecilia not dying would have made a big difference."

"Are you depressed?"

"If you could exorcise this, it would be amazing. Pharmaceutical intervention! Is that pure enough? Have you seen *Moscow on the Hudson?*"

"No."

"It's a good movie. Robin Williams? He plays a Russian sax player who defects in Bloomingdale's. I was just reading the *New*

Yorker review in your waiting room. Pauline Kael loved it: 'It's about going away forever, about not being able to go home.' "

"How's *your* writing going?" Dr. Band asked me, shifting in his worn brown leather club chair that matched his worn brown leather wing tips. Both the chair and the shoes could've used an Eric Anders shine.

"I'm writing, I just . . . When I was in high school Papi walked in on me one day while I was writing a poem. He looked like he'd just caught me masturbating. Not that I ever did."

"You never masturbated?"

"Mami and Baba Dora always told me that proper ladies always wear their underpants under their nightgowns. *They* did. Baba Dora suggested I keep my bra on, too, to keep my breasts intact and upright. I never had an orgasm until I was in college, and even then it was accidental. I had no *idea* your body could do that."

"That's rather amazing."

"It sure was news to me. But once I realized that I didn't need a partner to get one I was off and running. Anyway, my dad comes into my room and goes, 'What are you doing?' I go, 'Writing a poem.' He goes, 'Why?' He actually asked me *why*. He goes, 'You're going to starve to death with your little poems.' The poem got published in the Sidwell literary magazine, *The Quarterly*. Then Papi had it framed and put it in his office. It was about picking flowers with his parents in El Jardín Botánico in Miramar. They illustrated it with a black-and-white photograph of a little girl sitting alone on a picnic bench, looking far away."

"So in the end your father was proud of you."

"*After. After* it got published. Not before. I'm on my own here. Each keystroke types me farther away from them and the past. How are we gonna reconcile this?"

Dr. Band stared at me sadly and strangely in his worn tweed-ness.

"You won't be able to help me," I said. "I'm really sorry. I know you're trying but I think I'm gonna have to figure this out by my-self or maybe with a different doctor—no offense. I've been around shrinks my entire life. I can tell."

He sent me a final bill and a handwritten note on very fine sta-tionery, saying to keep in touch. My parents never paid. I guess they felt he didn't make me "well."

How do you explain to high-level WASPs who can use their plat-inum AmEx cards to pay for primo blow and weed, Ann Taylor sandals, psychedelic Peter Max outfits, and Harvard tuition how all-consumingly important it is to maintain your overall Juban-ique individuality by looking beautiful and having great nails? Well, mostly you can't. They'd laugh you off the lacrosse field. Or, in one particularly excruciating case, out of the music class. Poor Mr. William E. Fuhrman (B. Music, Catholic University). He just didn't know what to do with me or my frets issue. All of us had to play a musical instrument. I wasn't that interested, really, but I settled on the acoustic guitar; my parents played Segovia at their parties a lot and I loved the sound of classical Spanish guitar. So I got a beautiful curvaceous blond guitar on which to learn slightly more contemporary tunes, such as Neil Diamond's "Soli-tary Man." Jewish pop star, easy chords, what's not to like? Mr. Fuhrman, a seemingly mild-mannered sort, went around the room, inspecting our mouth or finger placement on our various instruments. My chord knowledge and strumming technique were decent but I was having trouble pressing down hard enough on the catgut strings to create the desired notes on the frets because my *leche evaporada*-fed veal cutlet fingers were so little and tender. The catgut always left painful grooves in them.

"I see the problem here," Mr. Fuhrman said, studying my left hand.

"Wrong instrument choice?" I asked. I was thinking maybe I should conduct, not play. After all, I was hell on wheels with a mascara wand.

"Wrong fingernail choice," he replied. "Why are your nails so long?"

"Neil Diamond, Paul Simon, the Beatles, some members of the Edison Lighthouse, Jimi Hendrix, James Taylor, Crosby-Stills-Nash-and-Young, the Stones, Bob Dylan, Johnny Cash, and Segovia all have long fingernails, excuse me. It's not a look *I* personally go for in men, but—"

"On their *strumming* hand. They have longer fingernails on their strumming hand."

"Uh-huh."

"Not on the hand they use to make the chords. Your fingering . . ."

"Uh-huh."

"You need to trim those nails."

"I drink Knox to grow them like my mom's. Cuban girls have to do it."

"Not Cuban girls who have to play the guitar."

"Them, too. It's *cultural*. And I don't *have* to play the guitar, by the way."

"You need to trim those nails."

"Uh-huh."

I could not say to him how important it was to Mami, and therefore to me, that I work on growing my nails. Nails that were infinitely more important in the long run to the chuppah than being able to play a friggin' guitar on an ordinary Wednesday afternoon. Mami had even gone all the way to the Safeway and bought me a whole orange box of Knox gelatin packets. The least I could do in return for such maternal sacrifice was to drink it. "Husbands like joo to have nice long nails," Mami had explained. "Joo can eh-stroke der back an' hair with dem an' look like a lady of leisure an' luxury with de beeg reenghs, like, from Teefany's, not like a fohkeengh peasant *sans* goo' jewelries." Mami had never once stroked Papi, but the *idea* was there.

Next music class my nails may have actually been a few nanocentimeters longer. The Knox was kicking in! But Mr. Fuhrman took me aside and we had another charming little tête-à-tête. "I'm not kidding," he admonished me. "I really mean it. Those nails have to get cut or there's no point in going on here."

"Uh-huh."

I kept right on with the Knox. Mami the Beautiful knows best. The class met again. Same thing happened. Only this time, Mr. Fuhrman didn't take me aside. He motioned me to come up to the front.

"Show me your hands," he said. I stuck out my arms. I noticed they were trembling.

"Spread your fingers wide," he said. He took hold of my left wrist to demonstrate it to the class. "This, this is unacceptable in a right-handed guitar player."

The class and I went silent. Then Mr. Fuhrman very calmly took a pair of men's nail clippers out of his pants pocket and clipped each of my Knox-grown fingernails, one by one. I heard them drop on the linoleum floor in tiny tinny hits. I felt hot. The roots of my hair were burning up like someone on fire. Juana de Arco. Hatuey. Anne Frank. *Gusano.*

"I warned you repeatedly," Mr. Fuhrman said, as blood rose up through my cheeks into my temples. I was sweating in my armpits and under my breasts and on the bridge of my nose, making my hideous eyeglasses slide down. My eyes got watery (thank God for waterproof mascara). I heard a few gasps and giggles. "Now sit DOWN," Mr. Fuhrman commanded, "and play that guitar PROPERLY."

My guitar may have gently wept for me, but if it did I never heard it because I never went near it again. I should have broken it over Mr. Fuhrman's head like Katherine did to her music teacher with a lute in *The Taming of the Shrew*. Anyway, the Knox stopped working after that. To Mami's infinite aesthetic disappointment and acute fear that I may never make her a grandmother, I resigned myself to having short, soft fingernails. I may never look beautiful and wear a Lucida diamond engagement ring from Tiffany. I may never be married and have babies. But at least I can type easy and not worry about chipping my polish, and I can do precise makeup application without worrying about poking out a Helen Keller eyeball, and whenever I cuddle Lilly I know I won't hurt her, and every time I touch the one I love I'm not afraid that I'll scratch his face or mangle the apprentice. That's got to count for something.

My small circle of Frenzy friends were people with an outsider feeling like mine whose intimate preoccupations were things like, *Am I too black? Are Angela Davis 'fros really okay here, or will they send me home to get a less overtly political Black Panther haircut? Am I too gay? Daddy hates me for being a fag and the locker room shower jock homophobes will hurt me if they see my red-painted toenails. Am I too fat? I'm gonna try the wheat germ–strawberry yogurt–cheddar cheese diet next and if that doesn't work I'm giving bulimia a whirl and if that doesn't work I'm going to Lourdes to bulk up my body on fabulous fattening*

French food so at least I can get credit for extracurricular activity to impress all my Ivy League admissions officers. Am I too transitory? What a D.C.-government cliché. Daddy may lose the next election and we'll have to go back to the relative sticks of Michigan/Pennsylvania/California.

My personal issue: WHAT THE FUCK WERE MY INSANE JUBAN PARENTS *THINKING?*

Our Southwest home address was deemed so über-gauche by Sidwell, so déclassé, that it was inaccessible to their school bus route. So I was sent to and returned from school each day by taxi, the other students greeting my arrival and departure by gathering at the windows to make fun of my four-hundred-pound driver—her nickname was Tiny. To brace myself for the daily humiliation, I'd dab on some extra Yardley lip gloss and tell myself I was a rich, famous, much envied Russian and Latina Jewess *princesa*-actress–prima ballerina from a banana republic whose potentate father had been illegitimately removed from power by a despot enemy's radical rebel communist army. Strictly as an interim measure, my Israeli supermodel mother and my dashing Latin king father had decided to deign to have me chauffeured to this school in a limo, but just until Papi regained his rightful rule and supremacy over our heinously misunderstood and very beautiful little native land. Denial may have been my parents' escapist drug of choice; mine was flights of imagination and, by and by, sex. Sometimes those tricks actually worked. Because every morning, it was the same horrible grind I'd have to rev myself up for, like Roy Scheider as Bob Fosse in *All That Jazz*, staring at his tired face in the bathroom mirror and saying, "It's showtime, folks!"

As I emerged daily from Tiny's taxi I considered telling her, "Tiny, let's get lost, let them send out alarms . . . Let's get crossed off everybody's list . . . And though they'll think us rather rude we'll tell the world we're in that crazy mood . . ."

Maybe Tiny, who lived in a trailer cahm with a skeletal unemployed chain-smoking cancer-ridden defeated husband named
Cletus, would also like to be a little runaway from the reality of
her life. We could just pack up the typewriter, Agua de Violetas
and TaB, get in that cab—and drive.

CHAPTER EIGHT

When Will I Become Like the Swallow?

In The Vagina Monologues, Eve Ensler writes that you cannot love a vagina unless you love hair. That is patently ridiculous—for a Jubana. That's like saying you cannot love an eyebrow, upper lip, jaw line, armpit, arm, knuckle, leg, or the top of your big toe unless you love hair. I love my body just fine, thanks (when I and a few choice others are not abusing or neglecting it), and I'm glad I have one, considering the alternative. But the only hair that I care to have on it is what's on my head, and even that grows as if it's on steroids, thereby forcing me into Jean-Paul's very expensive hair chair on a strict monthly basis. Growth means roots, of course. When Papi had surgery for a (benign, thank God) brain tumor in September 1999—it was the second such surgery in two consecutive years—my otherwise auburn hair turned white in a week. It's quite ugly and badly textured if left uncolored anymore, so I never do. Costs an unnatural fortune just to look Jubananatural.

I have a dark-haired, olive-skinned Latina heiress acquaintance who's had her entire body electrolysized from the hairline down—and even that was realigned.

"I used to be a gorilla," she said.

Well, once my puberty kicked in at age twelve, I never resembled a gorilla—I'm too Ashkenazi fair, thank God—but hair that didn't used to be there before suddenly was. And until I discovered shaving, tweezing, depilatories, waxing, and electrolysis for myself, that hair was there to stay, presumably even after I expired. The late Israeli poet Yehuda Amichai actually wrote a poem devoted to this unfortunately eternal grooming challenge, called "I've Grown Very Hairy": "I've grown very hairy all over my body. I'm afraid they're going to start hunting me for my fur."

Amichai started hunting me, and not merely for my fur, when I met him at a poetry reading in Washington in the early nineties. But that's another hairy story. For now, let's just leave it that he was one of my first older men.

Older men would become a leitmotif in my romantic life. Though I've been lucky to have a healthy father throughout my own life—until the second and crippling brain tumor surgery—I've also had a remote father. Once I hit puberty, that remoteness became physical. I remember seeing Papi at the end of the day and when I'd go to kiss and hug him, he was like a bull butting heads with a cow.

"Why did you ever have to grow up?" Papi would say. "You were so much cuter when you were little."

"How do you respond to that?" I'd ask Gramps, years later. "My father won't even touch me. It's like he wants to get it over with when he sees me. He's so uncomfortable. He won't sustain eye contact. It's like he's embarrassed to look at me."

"You hug him and kiss him and look at him anyway," Gramps

advised. "You take the lead, even if he doesn't respond at first. Believe me, you keep that up, he'll get the hang of it. You have to teach him how to show love."

"Why? Why do *I* always have to be the one to take charge of every fucking thing? Teach my own father. Why can't someone else be the mensch for a change?"

"Because," Gramps said, "you're stronger than the people around you. Certainly you're the strongest person in your whole fucking fucked-up family."

My whole fucking fucked-up family would have had a hysterectomy over that one. They think I'm a *pobrecita,* a poor little thing. I'm the least financially successful one of us. I'm the most observant and emotionally expressive. I'm the most creative and literary. I'm the most never-married and stuck with an alarmingly dwindling number of viable *huevos.* None of this holds any value. *Au contraire,* it creates *beaucoup de* hassle. What counts, what really matters in the Juban end? *Dinero.* Being attractive. *Dinero.* Never wearing the same outfit twice. *Dinero.* Marrying well. *Dinero.* Procreation. *Dinero.*

Otherwise you're a sohkehr. *You're going to starve to death with your little poems.*

As my very, very, very flush-from-nonkosher-pork-products brother Eric once remarked, most Virgo-ly, "In our, like, childhood, you were why we ever had problems and fights, okaaay?"

Okaaay! Was Eric saying this because I used to sadistically pull out his long eyelashes? Because I mimicked his strangely Valley Girlish diction and the way he made virtually every declarative statement sound, like, a question? Because whenever my parents went out on Saturday evenings and I was forced to babysit I'd throw him, baby Big Red Al, and our bulimic poodle, Martini, in the backseat of the car and take off barefoot in Mami's turquoise Corvair (garlanded with neon flower-power daisy stickers) on joyrides, smoking Mami's Kools and getting too mentho-latedly dizzy and phlegmy to figure out how to release the emergency brake?

"I may have been only twelve or thirteen but I was an excellent driver," I reminded Eric. "We never *once* got stopped all up and down Sixteenth Street and Military Road. And it was at *night*, too, when it's so much harder for a Helen Keller like me to see."

"You made Martini upchuck!" Eric said. "And then we had to clean it all up before Mom and Dad got home and spray it with Lysol?"

"Are you saying we *shouldn't* have sprayed it with Lysol?"

"That's not, like, the point or whatever?"

"It's *totally* the point. Name one Hispanic, regardless of class, who does not consider hygiene first and foremost. Even in Cuba, the poorest poorest Cubans, all they ever ask all the tourists for—if sex for American cash is a no-go—is SOAP, SOAP, and SOAP. *Jabón! Jabón* forever! Lysol would be, like, the Balenciaga of hygiene products for them. Hello!"

The conversation kind of devolved from there. No one in my family, least of all my brothers, ever has any idea what I'm talking about. Anyway, I'm sure Eric was right about everything because he is a pork products prodigy in a Roberto Cavalli leather jacket. And I am, after all, not.

I was twelve and Rebeca was giving me what would presently become my final nightly bath with Rebeca. As I got out of the tub I said. "I'm thinking of letting my hair grow out again."

She stared down between my legs and, smiling a weird smile, said, "You've got hair growing out all over now, don't you?"

What. A. Freak. That was the last time I allowed her or anyone else in my family to bathe me or see me naked. My new need for modesty and privacy made Mami mad, especially when I refused to get undressed in front of fifty strange women in Loehmann's cattle call dressing room in Rockville, Maryland.

"What ees de problem?" she asked. "Deyr not lookeengh at joo!"

"I don't like being stared at," I said. "And I don't like when *you* stare at me, either."

"Stare? I'm joor mother! I can see whatever I wan'!"

"Actually, no. You can't. News flash: This is *my* body."

"Dat came out of ME—an' den I had to say bye-bye to de bee-keenees an' hello maillots. Stretch marks. Really really beeg ones dat don' go 'way! Ever. Ees de same theengh eef joor room ees so messy. Eet can't be because dat room ees een *my* house! *¡El que paga, manda!*" Whoever has de money has de power!

"If you and Dad want to walk around in your underwear at home," I said, "that's your business. But I'm not. And I'm not gonna strip in this fucking dressing room. It reminds me of Auschwitz newsreels."

"No! Eet does NOT! Eet's not at ALL like dat! Dees ees *Loehmann's*, not de cahms! *¿Tu 'tás loca?* From where do joo get dees krehsee heestohree-ohneeks? Because ees not from ME, das for choor."

"Sunday school. That YOU send me to."

"Cahm Auschweetz ees a totally separate eeshoo dat has notheengh to do weeth de heres an' de nows of Rockveel, Maree-lan' an' Loehmann's. Joo can suffer a leetl beet in here because de prices are de rock's bottom. *Coño. ¿Qué te pasa a ti?* [Dammit. What is wrong with you?] Dees ees about de choppeengh an' de bargains, okay?—wheech ees probably de only goo' theengh about dees whole fohkeengh country—an' not about de gasses an' de chambers! *Por FAVOR. ¡Tranquilízate!*" P-LEASE. Calm down!

"I'll wait for you outside by the 'Deep, Deeper, Deepest discount' pile. Happy strip search."

Maybe those black-and-white Auschwitz newsreels really *were* too graphic for Sunday school sixth graders to see, the bulldozers sweeping up piles of naked, limp corpses. But at Temple Shalom, the Chevy Chase, Maryland, Reform shul we had joined in order to have no more "Chri-ist the Lord" episodes, students were exposed in small but unforgettable doses to Jewish realities, sometimes inadvertently, such as the fact that Passover candy sucks compared to Easter candy. (You can't beat the Reese's peanut butter egg with a jelly fruit slice. Sorry. You can't. I wrote a whole story about this in the spring of 1996 for the *Washington Post*'s food section. I offended Rabbi Bruce at the time, but I earned my Gentile candy cred.)

Temple Shalom was a cozy, modest, smallish congregation, with several hundred families. It had nowhere near the power, money, and clout of the better-known Washington Hebrew Congregation in Northwest D.C., where all the big-bucks Reform Jews belonged (including the few other Jewish kids at Sidwell). But Temple Shalom was a six-minute drive away from our new Neil Greene–designed house, located two doors down from his on a leafy, peaceful cul-de-sac of Neil Greene–designed houses, some of them with swimming pools in the basement, in Silver Spring. No swimming pool for us, but there was a turquoise bidet for Mami and lots and lots of wall space for Mami to hang all her suicidal Carroll Sockwell paintings and an atrium that Mami called "de hole een my house" from the roof skylight all the way down three floors to a boxed-in garden in the basement. From its top hung a cascading series of huge round glass light fixtures on white cords like an Alexander Calder mobile vertically turned on its side. Before dinner parties Papi would get a broom and pull in the cords to Windex *mis bolas,* my balls, as he called them. I used to tell my Sidwell friends that many, many men gave their lives to build the new manse, falling straight down the atrium shaft to their untimely deaths. A few gullible ones believed me. Sohkehrz.

I think Papi wanted to build Mami a castle, to re-create or re-capture or in any event approximate what we'd lost in Cuba after the fall and the expulsion from paradise, something grand, something amazing, something that could never be taken away. But the money ran out before the construction ended. This is because of my family's tenacious Cuban belief in dreams, magic, the fantasy of infinite resources. In reality, we were comfortable, not rich. That discrepancy, however, was at odds with Papi's need to always say yes, to Mami especially. It *seemed* as though nothing was beyond our reach and that we'd never ever have to really worry about money. That wasn't true, of course. But it is impossible to live within your means if your past life, a wistful, forlorn Cuban siren song, keeps beckoning and beguiling you like Circe. Americans say, "I think I'll re-create myself." Cubano exiles say, "I think I'll re-create my pre-Castro Cuban life."

It's like in *The Great Gatsby,* when, late at night, after another crazy party at his mansion, Gatsby tells Nick all that he expects from Daisy: to tell her husband, Tom, that she never loved him and to marry Gatsby:

"I wouldn't ask too much of her," [Nick] ventured. "You can't repeat the past."

"Can't repeat the past?" [Gatsby] cried incredulously. "Why of course you can!"

Baba Dora offered to make up the rest of the funds needed to finish the new house. She and Zeide Boris had been living with us in the guest room of the Southwest town house while Zeide was in the last stages of liver cancer. After our family's six-month Miami Beach layover in Las Casitas Verdes, Baba Dora and Zeide Boris and several of our cousins, uncles, and aunts settled in Charlotte, North Carolina, where there were plenty of textile jobs in the mills. It was hardly Camisetas Perro. The *Schindler's List* grayness of my grandfather's pre-Cuban life in Belarus had turned to blinding Technicolor in Cuba. Now after Castro, the grayness returned like *The Wizard of Oz* in reverse, Dorothy Galeowitz falling asleep in color and awakening to a black-and-white world. Sometimes we'd drive down I-95 South to my *abuelos'* modest Charlotte home and stay for a week, Mami's front seat Kool smoke blowing into my backseat face for seven straight hours, gagging me and intensifying my car sickness. My brothers and I would scramble out of the smoke-congested car and run around our grandparents' little house, Baba serving us orange-flavored Hi-C punch and ham sandwiches with French's mustard, melancholy Zeide smoking his cigars and having shots of *wiski* with Mami as the wet laundry flapped on the line in the backyard breeze.

In 1965 Zeide had developed cancer in his left eye. The eye was removed, and he wore a prosthetic one. The cancer eventually returned, though, this time to Zeide's liver. So he and Baba moved in with us for the last three months of his life. Mami came home from work every day at lunchtime to be with Zeide. She'd lie down next to him in his special rented hospital bed, and they'd talk. On the night of August 22, 1968, Zeide fell into a semicoma. My parents took him to Prince George's Hospital Center in über-gauche Cheverly, Maryland, where Papi had hospital privileges. I somehow felt Zeide would not make it because it was a Thursday. Thursdays in 1968 were horrendous: April 4, when Martin Luther King Jr. was assassinated and June 6, when Bobby Kennedy was killed. I remember going to see the Beatles' *Yellow Submarine,* and the tag line on the poster in the theater read, "The forces of good! The forces of evil!" I always thought of that whenever Gramps reminded me of *my* internal civil war.

"It's the conflict between light and darkness," he'd say, "between right and wrong. Imagine me sitting on your shoulder. If you ever give in to the bad, I'll slap you on the head and say. 'What the shit's going on?' "

"Oh," I'd say. *"That* sounds healthy."

And Gramps would say, "Thank you. It is. Sound mental health isn't about achieving some abstract human perfection. It's about being on to yourself, knowing yourself. So that when you're in a situation and you start to fuck things up the way you always do, you stop and go, 'Shit. I'm doing it again.' And then you can either stop it and self-correct or you can give in to it and knowingly, consciously, deliberately suffer the consequences. Now get the hell out of here. It's enough for me for one day already."

Zeide died during that Thursday night in the summer of 1968. He was seventy-four, younger than Papi is now.

"When we lef' de hospeetal very early in de morneengh," Mami recalls, "I felt like an orphan. I always felt notheengh really bad could ever happen to me as long as my father was alive. Den I look-ed at de sky an' saw a peenk cloud een de chape of an angel. I felt at peace."

As is customary in Latin families, Baba moved in with us, staying on for the next few years. She helped Papi pay for the rest of the house and spent her days in her southern wing, watching *Days of Our Lives*, sewing, occasionally baking her famous butter-marmalade cookies, and criticizing Rebeca's cooking and cleaning. Though Baba could be difficult sometimes, I felt protective of her because she was widowed and an innocent. When Mami took us out in the car—Baba always sat up front—I'd distract Baba from the backseat by leaning in between her and Mami and telling jokes and stories whenever we passed cemeteries. We were on Georgia Avenue in Northwest once, passing the Civil War–era Battleground National Cemetery. I saw granite soldiers and angels and I literally turned Baba's face away from them with my hands.

As we awaited completion of the new house, Ann "Nancy" Biester also came to live with us. There was nothing strange about or wrong with my Sidwell classmate, but we were friends anyway. Not soul friends, not a poor dead Cecilia-like sisterhood, but friends. Her Republican Congregationalist father, Edward G. Biester Jr., of Furlong, Bucks County, Pennsylvania, was up for reelection. In the fall of 1968 he went off to campaign for his seat as the U.S. Representative from Pennsylvania's Eighth District. So the blue-eyed dirty-blond Nancy moved in with us while her parents were away campaigning. I had twin beds in my room, and we'd stay up half the night talking, giggling, and fooling around. Rebeca was constantly barging in on us, checking to ensure we weren't turning into pubescent lesbians. After her father won the election, Nancy went back to her family, and our friendship ebbed. Edward Biester painted a watercolor of the Pennsylvania woods in winter and gave it to my parents as a thank-you for having sheltered, fed, and clothed his daughter for a semester. Mami hung the painting in the powder room of the new house, where it remains. We've never heard from the Biesters again.

The kids in my new neighborhood all went to Silver Spring public schools. The kids at Sidwell never set foot in Silver Spring, and really, rightly so. God knows I'd rather have been in the relative civilization of Friendship Heights or Chevy Chase or Bethesda myself. So that when Tiny pulled up at 3960 Thirty-seventh

Street, the Northwest middle school campus and, four years later for another four grueling years, at the upper school on 3825 Wisconsin Avenue, Northwest, she was taking me away from any after-school socializing and bonding possibilities, and delivering me back to an isolated cul-de-sac and a big house and Baba Dora sewing in the family room and Rebeca in the kitchen with her knives and cucumbers and Latina self-loathing.

"¡Usted siga con su tarea," Rebeca would tell me as I lowered a salted cucumber seed strip down into my mouth, *"y tal vez un día usted podra ser una secretaria importante!"* You keep up with your homework and perhaps one day you might become an important secretary!

Well, I *was* getting to be pretty good at typing on my Valerie Hermès typewriter with my lone right index finger.

The next morning was a Saturday. Rebeca always had the weekends off. Her daughter Beatriz would come pick her up and take her away to her crucifix-heavy Adams-Morgan apartment. I awoke in my bed in a pool of blood. My adorable lime-green-and-white-striped babydoll pj's were dripping, ruined. Thinking I must be dying, I screamed my head off like that Hollywood producer Jack Woltz with the horse's head in *The Godfather.* Mami and Papi rushed in. Papi took one look and rushed out. Mami said, *"¡Ay, gracias a Dios! Por tu madre,* joo can finally start dayteenth! Ees so great!"

"What the fuck are you talking about?" I said, looking down at my blood-soaked thighs. "Start dating? Dating who? I'm DYING over here, hello!"

Mami said I had just gotten my *período.*

"Das what de weemehns have to have een order to get to de nex' level, wheech ees de *tafetán color champán.*"

I asked her why.

"Dey johs do," she replied.

"And you get periods like this, too?" I said.

"I get dem," she replied, shaking her head, "but not like dat. Das, like, an extrehmeety. Must be from Papi's side of de fahmeely."

Mami put me in the shower, handed me a Kotex and an elastic belt with hooks, threw away my pretty pj's (this was all before they made Clorox Stain Out, my absolute favorite thing for such situations), changed my sheets, and called it a day.

Because I always had heavy and painful periods, periods of extrehmeety, especially Dreaded Day Two, I concluded that I was abnormal, since Mami's periods weren't at all like mine. Thus I was exceptionally moved one night in Baltimore years later, as another of my older men—in this case a Protestant—and I were getting ready for bed.

"I'm really sorry," I told the Geezer, "but I have my period. Today's Dreaded Day Two and . . ."

"Why are you apologizing for something that's natural?" the Geezer asked.

I had no answer. Or rather, I did, but it would take too long to explain.

"I have plenty of towels," the Geezer added. "And if you'd rather not make love, that's fine, too. I just want you to feel comfortable."

Voilà: The Geezer at his geeziest best. Gestures like those buy an otherwise impossible WASP man a whole lotta Jewish credit, a whole lotta Cubana time.

As worldly and precocious as I was in some ways, sexuality was still a mystery to me in the sixth grade, putting me at a huge disadvantage in the sex-crazed anomie of the sixties and to some extent the seventies. As for the eighties, well, I believe I was the only person *having* sex—with others, I mean—because Reagan was in office for most of that decade. Plus, if you recall, it was just a really, really depressing time: AIDS, poor dead Princess Diana marrying that awful Charles, shoulder pads, C. Everett Koop denouncing cigarette smoking, aerobics, Natalie Wood drowning, nouvelle cuisine, gravity-defying hair, yuppies, Salman Rushdie's fatwa, Madonna, George Bush. How horrifying is all that? What got me through—besides outrageous coitus (with an older married cowboy poet from Wyoming whom I called the Poet Lariat) in various hotel rooms with great room service across the nation—was, what always gets me through: TaB, Parliaments, and American pop culture. Americans are so *good* at diet drinks, cigarettes, and grooviness! Do you all appreciate what you have? Does it take a foreign-born Jubana to show and tell you how lucky you are and how good you have it? Seriously. When people start criticizing this country I go krehsee for two key reasons: One, I just don't hear about people dying to get into, say, oh, I don't know, how about . . . CUBA! Really, when was the last time you heard about any illegal immigrant risking his or her life to go live in CUBA! And two, how can you criticize America, a country that (a) lets you criticize it (something no Cuban living in Cuba can do), (b) has TaB (something no Cuban living in Cuba can have), and (c) has had, even in the politically tacky eighties, the following (which no Cuban living in Cuba has probably ever even heard of, considering that Cuba's stuck in such a pathetic time warp that acid-washed jeans are cutting-edge): the TV shows *Hill Street Blues* and *Cheers;* the movies *The Accused; Amadeus; Atlantic City; Babette's Feast; Baby Boom; Blow Out; Blue Velvet; Body Heat; Casualties of War; Cat People; Children of a Lesser God; Dangerous Liaisons; Desperately Seeking Susan, Diner; Down and Out in Beverly Hills; The Elephant Man; E.T.; Fatal Attraction; Full Metal Jacket; Hannah and Her Sisters; Kiss of the Spider Woman; The Last Emperor; Married to the Mob; Mommie Dearest; My Beautiful*

Laundrette; My Left Foot; New York Stories; 9 1/2 Weeks; Out of Africa; Prick Up Your Ears; Prizzi's Honor; Raging Bull; Raising Arizona; Reds; Risky Business; Roxanne; Scarface; sex, lies, and videotape; She's Gotta Have It; The Shining; Sophie's Choice; Steel Magnolias; Tequila Sunrise; Terms of Endearment; Tootsie; Trading Places; The Untouchables; Wall Street; The Witches of Eastwick, and *Working Girl;* and the music of Anita Baker, Edie Brickell & the New Bohemians, Eric Clapton, Fleetwood Mac, Joe Jackson, Michael Jackson (prefreakdom, obviously), Rickie Lee Jones, the oft-overlooked Katrina & the Waves, John Cougar Mellencamp, the Police, the Pretenders, Bruce Springsteen, Steely Dan, U2, Suzanne Vega, and Stevie Wonder.

So let Mami call this nation *una mierda* and let that cheesy Puerto Rican Bernardo in *West Side Story* sing "Everywhere grime in America, organized crime in America, terrible time in America." As his hot little Puerto Rican girlfriend Anita replies, "Joo forget I'M een Amehreeca!"

Thaaat's right. Okay. End of jingo jingle.

As I was saying, I don't know about other Hispanic parents, but mine certainly never discussed sex with me, nor did they ever exhibit the typical overprotection that most Hispanic parents— fathers, especially—do over their daughters. I really can't say why, and I doubt the 'rents could tell you, either. Though my father is a medical doctor and my mother is a psychiatric social worker, all their professional training was useless at home. Irrelevant, even. The only thing I ever heard either one say about sex was once when Papi said, "A woman should be a lady in the living room, a chef in the kitchen, and a whore in the bedroom."

I asked what a woman should be in her home office, and Papi was, like, *"¿Qué?"* Say whuuut?

• • •

Poor Mr. Kevin J. Kinsella (S.B., MIT; M.S., Johns Hopkins University). The boyish brunette was assigned to teach us Sidwellian sixth graders about human reproduction. We were sitting at long tables in the biology lab as he calmly explained about sperm and semen and eggs. I couldn't help it, I just burst out laughing.

"WHAT?" I cried. "No way! My Cuban refugee parents do NOT do that!"

Everybody laughed. The petite cool girl sitting next to me, Theresa Rosenblatt, whispered, "Jesus, don't exaggerate so much. You want to pull his leg, fine. But you're overacting."

I wanted to tell her that there was no way I could *over*act because I was a bona fide drama student and I knew the difference between drama and melodrama (although granted, that is a toughie for most Cubans to distinguish). I wanted to tell her that I had not been feigning sexual ignorance, that I actually *was* sexually ignorant, and that all of this information was coming as a real shock. I wanted to tell her how much I hated her and her mean little affluent cool white *gringa* girl gang clique. I wanted to tell her how much I wanted to be her or to be them or to be Twiggy or Nancy Sinatra or Mia Farrow or Ali MacGraw or Lauren Hutton or Julie Christie or Faye Dunaway or anybody but myself.

But I didn't.

I didn't utter a single word. ME. All I felt was the same sensation I'd had back in that awful guitar class when the teacher cut off my Knox-inspired nails. Blood rising up through my cheeks into my temples. Sweating in my armpits and under my breasts and on the bridge of my nose. My hideous eyeglasses sliding down (new but no less hideous ones: heavy dark un-ironic tortoiseshell cat-eyes): *Quando fiam uti chelidon/Ut tacere desinam.* When will I become like the swallow/That I may cease to be silent.

Sidwell Friends. Sidwell Enemies. The school, the people in it, the entire atmosphere, was trying to silence me under a sea of embroidered whales and Lacoste crocodiles and their respective spawn. Silencing me slowly, gradually, like a piecemeal Jubana spirit kill.

As I left the lab, a fringe member of the Theresa group strode over to me: Peggy Loomis, a great big tall, strapping, swarthy, black-haired, black-eyed giantess with an equally gigantic and long face. She was into field hockey and non sequiturs.

"That was funny what you said in there," Peggy said, looking down on me. "Your arms are really hairy."

"Thank you," I said.

"I mean they're blond, but still."

"Gee, thanks," I said.

"Ever seen *The Sound of Music*?"

I had no idea what this had to do with my hairy blond arms or what my hairy blond arms had to do with anything in the first place, but I went along with it. Jubanas can be too trusting of other people's actual motives. Isn't that what happened to us with Fidel Castro? Do we ever learn? Uh, no.

"I loved *The Sound of Music*," I told Peggy. "And *Mad* magazine, they did a spoof of it that was so funny! *The Sound of Money.* 'The hills are alive with the sound of money . . .' Love *Mad*. But my mom really hated the movie. She said Hitler was born in Austria and the Austrians were some of the most vicious, virulent Jew haters. So we hate Austria. We love the nightgowns, though."

"What?" Peggy said.

"Lanz of Salzburg. They make the best flannel nightgowns for winter. Tyrolean prints are so cozy and sweet and toasty. Nerdy but fabulous. My mom couldn't believe *The Sound of Music* movie, though. She goes, 'Dat Maria ees so fake. Nobody can be dat goo.'"

Peggy began singing: "'Edelweiss, Edelweiss/Every morning

you greet me/Small and white, clean and bright/You look happy to meet me.' "

In spite of that raging Austrian anti-Semitism, I joined right in. I really love that song. And I've always been good at memorizing words and lyrics.

" 'BLO-ssom of snow, may you bloom and grow,' " we continued.

I was beginning to feel better. Music can change everything, especially mood and outlook. Try staying depressed while listening to, say, "Here Comes the Sun," one of my all-time fave songs. Peggy's black eyes planted themselves on where she approximated my nipples must be. My breasts, like the rest of my body, were becoming Cubanly voluptuous, the absolute worst kind of body type you could ever have at Sidwell Enemies. Sidwellemies. I was so trying to fit in, if only sartorially, and it was so not working. Petite busty girls with curvaceous derrières look absurd in asexual L. L. Bean or Brooks Brothers striped rugby pullovers and flat-butt straight-leg khakis or thin-wale cords and Earth shoes, combat boots or Clarks Wallabee suede ankle moccasin desert boots—a Sidwell female's idea of stylish, feminine dressing. (Fridays were casual, but still heinously conservative. We weren't even allowed to wear blue jeans for the first few years.) Peggy placed her hands in front of her own nonexistent chest and began making outward circles, suggesting big growing breasts.

"Bloom and grow forever," she continued singing. She was laughing. Her symmetrical teeth were gigantic white peppermint Chiclets. I had ugly braces on my teeth. Like my frizz-prone hair (amusing new hormones had altered its previously straight, glossy texture); astigmatism; myopia; flan-size pores; acne-prone and perpetually ruddy—in the T-zone—oily complexion (same amusing new hormones struck); soft, short nails; and flat, wide feet, Mami blamed my acute dental sensitivity and need for orthodontia on Papi and the Andurskys. My orthodontist turned out to be ineffectual—my teeth actually worsened after his treatment, causing me to have to get braces all over again as an adult—but my parents kept sending me to him because he was a distant Jewish relative and his office was three minutes away from our house.

" 'Edelweiss, Edelweiss,' " Peggy derisively sang on, solo. She was sniggering and laughing so hard she could barely snort out the rest of the words. " 'Bless my homeland forever.' "

I'd long since stopped singing. It was happening again: The blood ascending under my cheeks up through my temples, the streaky squeamy sweaty wet all over. I felt myself melting. Peggy was one wicked Baroness Elsa Schraeder. To take a perfectly innocent, sweet song about a flower and perversify it into a mean little put-down over something I had no control over . . . I stood there for a minute, watching Peggy stride away confidently, her mile-wide North American shoulders still shaking from her laughter.

This was some kind of country. You had to assert yourself constantly: Kill or be killed. It was exhausting. Was life in the United States just one long Western shoot-out? Because it sure as hell wasn't my tradition. Jews and Cubans really aren't into that Big John–High Chaparral sort of thing. So instead I thought about the story of David and Goliath. Did David *like* having to deal with Goliath? No. But he did. He had to. And it may have happened three thousand years ago in the Middle East, but I felt David's chutzpah and courage was applicable to me in the sixth grade in 1969 in Washington, D.C. It was then my Jubanite BTT (Bullshit Tolerance Threshold) kicked in. It tripped the alarm in my brain, making it beep.

"Hey Peggy!" I said, running to catch up with her. "If you're jealous 'cause I've got great big fabulous Cuban bazoomies and you have, like, none, maybe you need to see somebody. My mom could get you a shrink. My dad could recommend a plastic surgeon. Board certified. And now, as the baroness would say, '*Auf wiedersehen,* darling.' "

Peggy Loomis stayed far far far away from me after that. One maroon and gray *gringa* Goliath sohkehr down, seventy-plus to go.

CHAPTER NINE

Tricky Dick

What marks a rite of passage? In my own life, my Bat Mitzvah, for one. What was supposed to be a profoundly serious and moving event—I confirmed my religion before I learned geometry—was recorded for posterity but no record exists. (And I'm not just saying that because my little friends and I snuck into my bedroom afterward for a post–rite of passage passage, i.e., smoking pot for the first time, and I was hazy.)

I'll explain. Stay with me here, you need context. Catholic Latinas have a rite of passage ceremony on their fifteenth birthday called the *quinceañera,* or simply *los quince.* You wear a massive dress resembling a lace parade float, your parents go into massive debt for the ensuing fiesta, and you get a massive number of gifts. Many of those gifts will not be cash or checks and will therefore be disappointing. But! Your well brought-up mami will make you send enthusiastic hand-written thank-you notes to everybody. Expressing appreciation is what civilized people do. It smacks of maturity. As the old saying goes, Today you are a woman. Tomorrow it's back to eighth grade.

For pre-Castro Jewesses, however, there was no correspon-ding ritual. (Just as there wasn't and isn't for Catholic Hispanic boys. This must be because all such boys are born grown-up macho men.) Sure, Jubano boys all had Bar Mitzvahs at thirteen, as Papi did. Traditional Judaism, like all traditional religions, is patriarchal. Some may even say misogynistic, but let's not go there. Anyhow, there was no Reform Judaism in pre-Castro Cuber. You had Conservative (my family belonged to El Patronato de la Casa de la Comunidad Hebrea de Cuba), and you had Ortho-dox, and neither recognized a girl's coming-of-age. So Jubanas co-opted the *quinceañero.* Any excuse to shop for cute new outfits.

"Beauteefool party een our home," Mami recalls of her own *quince.* "My long dress was light blue embroider-ed organza. I have a peecture."

Too bad I don't. Not a single one. Of my Saturday, May 22, 1971, Bat Mitzvah, I mean. Not that I'm bitter. It was only the most important day in my young Jewish life. I had only spent six intensive months every day after school at the Hebrew teacher's house, preparing my haftorah and crash-learning the Other mother tongue. We had only invited one hundred friends and rel-atives from across the country to the ceremony at Temple Shalom and to the lunch party catered in the new house's expansive backyard. Plus, I had to pay, albeit at a discount, a klepto Sid-wellemy consort for a huge eye shadow palette I coveted. You'd see makeup you wanted at the store and tell her, and she'd pro-duce it within a week for half of the retail price. This particular item was a big plastic turquoise compact containing a virtual rainbow of assorted pastel hues, most of which picked up the col-ors of my bespoke Bat Mitzvah dress. In 1971 so-called granny dresses were all the rage, sort of a prairie-Victorian attitude. You'd wear them near ankle-length with kick-ass construction or cowboy boots. Mami bought the fabric, a peach, ivory, and violet pattern of tiny roses inside vertical stripes that looked like a Victorian girl's bedroom wallpaper. We took it to Mercedes, a Cuban friend and talented seamstress who lived in Virginia. Mercedes could make anything. She once made me a wonderful ice-blue linen coat dress with faceted navy-blue buttons like Julie Christie's in *Darling*. And Mercedes made me my Bat Mitzvah dress, which I wore not with boots but with a pair of strappy low-heeled white sandals in size 7N that killed the PVCs. I spent an hour in the bathroom coordinating the makeup with the outfit.

Not that there's a single picture of it.

Mami's older brother, Tio Jaime, *quien en paz descanse* (may he rest in peace), was an amateur shutterbug. He offered to shoot the event as his gift. Ever eager to save a buck—on me, especially—Mami agreed at once. So Tio Jaime shot away, hundreds and hundreds of pix.

Not a single one came out.

Tio Jaime forgot to remove the lens cap.

It was a pivotal you-get-what-you-pay-for moment, which is why, when it was Eric's and Big Red Al's turns, Mami had professional photographers shoot their respective Bar Mitzvahs. Which is why I'm glad I was stoned for most of my own party, not that I particularly enjoy grass or similar major hallucinogens. But on that day it did kind of prepare me for the inevitable subsequent fallout. A week or two later Mami "mentioned" it as an aside, avoiding direct eye contact. I'd just returned from seeing *Klute*, however, and though I lacked the requisite don't-FUCK-with-me shag haircut and tiny pert breasts with protruding nipples, my Bree Daniels-prostitutally-inspired bullshit detector was on high alert. I was thinking, *What would Bree do? What would she say?*

"Joo know, I'm really sorry about the peectures, *mumita*," Mami said. We were in her bedroom. As usual, the TV was on, loud. Martini lay curled asleep on the bed, a pink satin ribbon tied in her manicured head. Three hundred and ninety-seven *Hola!* and *Vanidades* magazines were stacked on the floor. Mami was sitting in a rocking chair facing the TV, surrounded by three picnic baskets containing seventeen thousand mani- and pedicure accoutrements, a lit Kool in a stolen Argentinean beefsteak restaurant ashtray, and an espresso. The combined smell of cigarette smoke, acetone, and foamy Cuban coffee permeated the air, pungent and powerful.

"Pictures?" I said in my best ironic, sardonic, world-weary Bree Daniels imitation. "What pictures? There ARE no pictures."

"Well . . ."

"Way to go, *mother,*" I said. "Jesus Christ, think you can open a window? The fumes in here could kill Martini."

Mami shot me The Look. I stared right back at her, my arms crossed.

"Joor oncle meant well," she said. "What can joo do? Cheet happens. Das eet. De meelk ees on de floors, speelt."

"Thanks for the memories, *mother,*" I said. "I can see why it was all worth it now. Maybe we can get Tio Jaime to shoot Eric and Alec's Bar Mitzvahs. *Mother.*"

I knew I was being insufferably bratty—which is perhaps why Papi preferred me prepubescent—but this is what hormones do to you at this age, so I couldn't help it. Mami may have felt ten times worse than I did, and was probably embarrassed and guilty as sin, too. But I didn't give a shit. She should feel bad. I saw the whole ignominy as her fault for choosing cheapness over noncheapness and for not having used better judgment. At least I'd gotten a few good gifts out of it: Anitica, Mami's Cuban friend who had sat with her in silence in Mami's locked Miramar bedroom after Cecilia's death, had given me a beautiful antique rose gold bracelet with an opal and two tiny pearls. Nedda, the wealthy bohemian Cubana who lived in Mexico City and whom

Papi briefly dated in Cuba after a spat with Mami, had my invitation framed in an elegant hand-painted rustic Mexican frame. My Sidwell friend Laura Hart, the Lauren Bacall-ish daughter of the late Senator Philip Hart (as in Hart Senate Office Building), the distinguished Democrat from Michigan who was the only senator in his time to sport facial hair, gave me a wonderful photo album with a psychedelic pink and yellow hologram cover. Not that I'd have any B-M pix to put in it. Oh well. I could always go drown my sorrows in Laura's sunken round bathtub. Every time I'd go to sleep over I got to bathe in it. Laura would turn on the jets and sit on the rim while I'd have myself a lovely whirlpool. Plus, we never used the same towels twice. Laura's maid changed the towels and bed linens every day at that house, and not just for guests. I had no idea American people lived this way. It was so exciting.

"Joo know what?" Mami finally said, exhaling her smoke in my face in a streaming cloudy bullet. "I hope dat when joo have a daughter chee comes out johs like joo. Den an' only den weel joo know what I have to deal weeth."

"Poor you," I said. "You have it sooo hard. I'll bet you were just a dream angel of goodness child when you were my age. I'm sure that's what Baba Dora would say. *Mother.*"

The only reason I was able to get away with that snottiness and retain my dentiture was that Mami's lap was covered with a hardback book that was covered with a paper towel that was covered with nail polish and acetone bottles, orange sticks, and cotton puffs. Also, her nails were wet. When it's a choice between smacking your provocative pubescent upside the head and not fucking up your nails, Mama Jubanas have their priorities in order.

• • •

Life in general and Algebra I in particular and I were *not* getting along. Cutting class, shopping, and *The Mary Tyler Moore Show* were pretty much all I had to live for. I was moody, restless, unfocused, a Jubana out of water. Nobody seemed to notice or care. *That's how teenagers are.* When I'd pile up enough demerits to get suspended for, say, talking too much in class, smoking in the bathroom, or wearing too much eyeshadow and too little minis (my fave that year was a faux pony hair micro-mini), I'd think of it as an opportunity to catch up on lost *Days (of Our Lives),* sleep, and back issues of *Seventeen, Rona Barrett's Hollywood,* and the Style section, sort of a little tasty-break from daily hell. Mami would pick me up at the unamused headmaster's office and scream obscenities all the way home as we barreled up Wisconsin Avenue and through Rock Creek Park's circuitous single lanes. She once drove so fast on Wise Road and Beach Drive, she almost ran over a Bambi!

"You practically hit that baby deer!" I yelled as we screeched past yellow-and-black horse-rider and deer-crossing road signs. "Animals live in the park! Are ya gonna try for a pony next?"

"Fohk dem," Mami said, exhaling her Kool smoke. "Deyr not endanger-ed. But joo are. Oh an' by de way, CHOHT *UP.* Joo are grounded. Joo don' get to talk."

"Bitch, I can smell your tires burning," I said.

"Well guess what?" Mami said. "CHOHT *UP."*

To teach me a REAL lesson, Mami hit me where it would hurt most: She confiscated my fabulous stolen eyeshadow compact, the one whose pastel palette I'd used to coordinate with my Bat Mitzvah dress. That was cruel. But, whatever. The suspensions only temporarily suspended my *amour fou* for talking, nicotine, makeup (I was amassing a private stockpile), and anti-Quaker haute couture. In terms of a fear factor, punitive techniques didn't work; they only made me more mutinous. I'd cut more classes, feel madder at the punishers and more aloof about consequences. Mrs. Katharine "Kate" Henry (B.A., University of Reading; L. es L., University of Dijon), my veddy British English teacher, once observed, "Gigi, you are so veddy unbalanced!" I said, "Indeed, m'lady, I quite agree with thee. But what do we DO about it?" She just shook her hoary Dickensian head.

There it was: Observation astute, solution nonexistent. Concept: Since my intellect wasn't at issue, how about exploring other possibilities, like, oh I don't know, how about . . . MAYBE THE JUBANA-ANASTASIA REFUGEE PRINCESS IS FUCKING DEPRESSED IN HER HOME AND SCHOOL ENVIRONMENTS AND NEEDS TO SEE SOMEONE AND POSSIBLY MAKE SOME CHANGES. Maybe something along the lines of a really good adolescent psychiatrist. I'd have never been in this mess in the first place if only those black parents in Southwest had adopted me as I begged, cajoled, and wheedled them to. But no. And the therapeutic support option just never occurred to anyone. Sorry, but you can't expect a kid to think of everything, even a kid like *moi*— that's what teachers and parents are supposedly for. Back in the dark ages of 1972, though, depression was stigmatized. I remember feeling more depressed than usual when George McGovern dumped his Democratic presidential running mate Thomas Eagleton (whose sweet towheaded son Terry was a schoolmate and friend) because he'd had electric shock therapy and had been hospitalized for depression and bipolar disorder. That July I was on Tío Bernardo's motorboat. As we bounced through the choppy, foamy blue waters of Biscayne Bay, Tío kept hollering, at no one in particular, "MAHK-GOH-VERN! MAHK-GOH-VERN!"— that is, until MAHK-GOH-VERN made Eagleton VAH-MOOS. (Tío's battled his own depressive demons, so who could blame him?) During Jimmy Carter's administration, Tío Nano was Carter's lead negotiator in *el diálogo*, the dialogue, with Fidel Cas-

tro to free thirty-six hundred political prisoners. Tío made fifty secret trips to and from Cuber to discuss the deets with El Caballo,
whom Tío considered a despicable asshole. Hence, no standing
on ceremony. Cubans rarely do. Tío Nano remembers it vividly:

FIDEL CASTRO: (using the formal tense, *usted*): *¿Cómo está, Benes?*
TÍO NANO: (using the familiar tense, *tu*): *¿Cómo estás, Fidel?*
F.C.: *¿Qué?*
T.N.: Free the thirty-six hundred political prisoners. Let them re-
unite with their families in Miami.
F.C.: *¿Qué?*
T.N.: You don't give a shit about them, so do it. Besides, it will
make people admire you. Like a pharaoh.
F.C.: *¿Qué?*
T.N.: Moses is my hero. I align myself with him. You once said,
"History will absolve me." Prove it.
F.C.: *¿Qué?*
T.N.: And while we're at it, you owe me one million American dol-
lars.
F.C.: *¿¡QUÉ?!?*
T.N.: That was the net worth of my father Boris Benes's business,
Camisetas Perro, when you stole it in 1960. Oh, and I'll take a
check, although I'd prefer cash. Then we'll understand each
other better.
F. C.: *¡Adios, cabrón!*

Castro never did cough up Zeide's purloined business—*quelle
surprise*—but he did the prisoners, who were tearfully welcomed
back into the arms of their Miami kin. You'd think the Cuban
Americans would all be grateful, right? Oh please. Tío Nano's life
was practically ruined. He had so many death threats by fellow
Cuban Americans that he took to wearing a bulletproof vest to

work every day for years—while he still had a job to go to, that is—
and was accompanied by a bodyguard. Tía Ricky was at risk. The
kids—my cousins Joel, Lishka, and Edgar—were derided at
school every day. All because certain members of the exile community could not "forgive" my uncle for committing the ultimate
exile sin: talking to the dictator. I'm sorry, but that is fucked up.
And it hurt me, too. Whenever the *Post* sent me down to Miami on
assignment, I could never meet publicly with Tío Nano—on whom
I relied for background and sources—because he was such a bête
noire.

"I'm persona non grata here," Tío told me. "Eef dey see us at
Versailles [restaurant], nobody weel talk to joo. Deyl theenk joor
een bed weeth de dehveel. Joor life might even be een danger."

We stuck to phoners after that.

Hence, back at Frenzy and barring an expander, I'd just hop the
30 bus and go down Wisconsin Avenue to Georgetown for lunch
at Maison des Crêpes, which I fondly called Maison des Craps,
and hot pink and lime-green espadrille perusal at Pappagallo
(very Palm Beach)—or up Wisconsin Avenue to Friendship
Heights to grab a bite at Booeymonger's for preshopping suste-
nance and then hit Woodies department store. On those days I'd
instruct Tiny in the morning to pick me up wherever I was on that
given afternoon. She always went along with it. Sometimes my
platonic Frenzy boyfriends Rob and Peter would join me on those
extracurricular excursions, and if we had the money we'd eat
chef's salads with gallons of Thousand Island dressing and sweet
little corncakes with butter on the side at HH (Hamburger Ham-
let). Then I'd hit Woodies, with or without them. My father had
given me a credit card "for emergencies only," which had to be a
joke since I didn't even have a legal driver's license yet (not that

that had ever stopped me from taking my illicit joyrides), and Papi's technical definition of "emergencies" precluded anything short of engine failure. *My* "emergencies," however, were pretty much all retail. Mami didn't buy me enough clothes, that was all there was to it. She didn't enjoy going shopping with me because we always ended up fighting. Our styles and approaches were incompatible. She was Loehmann's, sales, quantity, earth colors, amassing. I was Ann Taylor, full-price, high-quality, happy colors, minimalist (not really, but at least my nickname was never Imelda).

For his part, Papi would do anything to preserve the domestic peace and status quo and avoid any confrontations whatsoever. So when he reviewed the credit card statements every month he never said a word, even if I had gone berserk. My semierroneous interpretation: There were no limits to what we could afford; I could only ever depend on Papi, not Mami, for financial aid; Papi felt guilty for doting on and preferring my brothers, whom he regarded with moist-eyed hero worship, so allowing me to charge myself into a Jubanique stupor on his dime was his way of paying me back for that lack (although my routinely doing so would advance and deepen his resentment toward me for "using" him); having a budget, sticking to it, and openly discussing money is boring, taboo, threatening, and vulgar; life has no real cost if you have Papi for a dad; and it will all go on and on like this eternally because unlike mere mortals, Jubanos are too Ju-ttractive and Ju-special to ever get sick, incapacitated, old, or die. Max out one credit card, open a new one.

Repeat.

Forever.

It was a rude shock—to Mami as much as to me—to discover how brief Forever is. After Papi's second surgery for a brain tumor, when Mami took over the family finances, we realized how much strain we'd both put on Papi's bank balance.

We'd discuss at length the relative merits of Herbal Essence shampoo over Flex (Rob and Peter both had long, straight, thick hair that retained the shampoos' smells better than mine; my Agua de Violetas may have contributed), and Rob dished the dirt on Cher, his idol, with his customary Robin Williams manic delivery. Peter regaled me with stories of weekends at Camp David, where he and Rob went with Peter's dad, H. R. Haldeman, Peter's mom, Joanne, known as Jo, and President Nixon. (We never discussed politics. I don't ever remember Peter saying "Watergate.")

Jo always told the boys, "You cannot go to Nixon's cabin, there's Secret Service all around it." Now, Peter was brilliant, hysterically funny, and gorgeous. He had probably the prettiest face I'd ever seen on a boy. Huge hazel eyes with very thick, long, black eyelashes, a peaches-and-cream complexion, and that perfect, naturally streaked blond-brunette hair. He could have been a Leonardo *cherubino*. But Peter, who favored navy-blue Chuck Taylor All-Star Converse canvas oxford low-rise sneakers, was a provocateur with an edge and really strange mood swings. This one time, Peter talked Rob (who could be talked into anything, a key part of his charm), into a golf cart in the middle of the night. Peter started driving a thousand miles an hour to the presidential cabin; everybody stayed in separate cabins, whereas I'd always imagined Camp David as a kind of ski lodge situation. He told a terrified Rob to "just follow my lead, darling." They charged right up to Nixon's cabin and sirens went off and two dozen guns were suddenly trained on their heads.

"WHAT ARE YOU DOING HERE?" a security agent barked.

"I'm Peter Haldeman," Peter said, nonchalantly. "H. R. Haldeman is my father. We're here to see the president."

"YOU CAN'T MEET HIM!" the guy said.

"Sir, I really feel *terrible* about bothering you at this late hour," Peter continued, dabbing faux tears and sniffles away, "but my friend here, he doesn't have much time to live. [Whispering.] Terminal cancer. [Pause.] And all he's ever wanted is to see where the president of the United States stays in Camp David. [Whispering.] Dying wish."

"Sorry about your friend. Really. BUT YOU CAN'T MEET THE PRESIDENT!"

In the morning, Jo confronted Peter, who actually took pleasure in seeing if he could get out of these things: "It was late, Mother. We got lost. I could have SWORN it was our cabin. I got mixed up and turned around. I feel just *terrible* about compromising dad." And H.R., who was one scary mutha to begin with—that buzz cut alone—would EXPLODE. Peter knew this was not proper. Going to Nixon's cabin during the day would've made more sense. But Peter was just like that.

Another time there, Peter and Jo were playing doubles against H.R. and Rob. There was a telephone on the tennis court. It rang and H.R. told Peter to pick it up since it was on Peter's side.

"It's DICK for dad!" Peter shrieked, not covering the receiver. "Dad, it's DICK. DICK wants you—NOW!"

Jo fainted, Rob bit his lip, H.R. sighed, and Peter threw his head back and guffawed his trademark "AH-hahahahaha."

Peter left Frenzy in April 1973, tenth grade, one month before H.R. was forced to resign over Watergate, one year before H.R. was indicted, not quite two years before H.R. was convicted and sent to a federal slam for eighteen months. The Haldemans moved back to California, where they were from. Nobody at school knew what really happened. I saw Peter and his parents in the maroon and gray upper-school hallway by the headmaster's office, all looking very grave, and that was it. There were lots of rumors. I got a card from Peter later on, with a color photograph of him with H.R. (buzz cut grown out), Jo, and Peter's three attractive siblings, Susan, Hank, and Ann. Except for H.R.—who's off to one side inspecting what appears to be a camera, slide projector, or tape recorder—the family's standing around a yellow touring bike in the yard of the Haldemans' house. Peter's leaning on the upright bike, looking down at Hank, who's on the ground checking the rear tire. The environs were ordinary yet lush, a lush life "where one relaxes on the axis of the wheel of life / To get the feel of life . . ." Three pots of red chrysanthemums are lining the brick porch, with lots of sun shining on the flora. I wrote Peter back but we lost touch after that. I always adored him, though. At Frenzy, Peter made me look normal.

After recording the day's events on my Valerie typewriter, I'd get in my pajammies, sprinkle Agua de Violetas in my hair, refresh my body with Yves Saint Laurent Rive Gauche Eau de Toilette spray, and pore covetously over my latest FBS catalogue—my kind of nighttime poetry and porn—with the pages folded into neat triangles on the outfits I desired. I always got tons of fashion catalogues in the mail but I really lived for FBS. The letters stood for French Boot Shop, an upscale, hyper-trendy boutique in exotic, far-flung New Rochelle, New York. (That's where I bought my Corkies, those fabulous pigskin crisscross sandals with a cork wedge heel that I scuffed on that psychotic plane ride from Mexico City to Oaxaca with the 'rents when I was sixteen.) If the *Sex and the City* gals were catalogue shopping in the seventies and eighties, FBS would have been their fashion mecca. (Of course, they would have lived in Manhattan, so why would they have bothered?)

Living in Washington, D.C., an excruciatingly conservative, government- and politics-obsessed, culture- and style-starved city of streets named after states and granite monuments and mausoleums, presented serious sartorial challenges. So I was forced to shop outside the federal box, as it were. FBS transcended my fashion magazines because unlike the mags, you could actually order the clothes in the photographs. FBS showed me the girlishly unexpected and lavishly quirky that somehow just worked: a mustard-yellow Norma Kamali fleece pullover dress in the shape of an inverted triangle with an exaggerated funnel turtleneck and Velcro'd shoulder pads, worn with opaque black tights and red suede Joan & David pumps. A poofy Esprit khaki flight suit feminized at the waist with a wide cinched floral embroidered belt, and worn *sans* socks and with copper metallic Joan & David flats. A lemon-yellow Betsey Johnson halter tube dress scattered with tiny pink rosebuds, with a shirred, gathered bodice and a full skirt, worn with high-heeled wooden slides whose straps were clear plastic panels decorated with lemon-yellow and rosebud-pink sequined butterflies.

Those images, those wonderful clothes, most of which I could never afford (Papi didn't give me a hard figure limit but I sort of intuited how far I could push it) or wear to school or anywhere else I was likely to go, conjured up correspondingly fantastic situations. I'd cut out the images and glue them to the back of my bedroom door, creating an intricately layered, overlapping collage, and stare at them. That was one way to acquire the attire. Pin down the outfit and the occasion will present itself, that's my motto. For instance, if I only owned that icy pale blue cashmere bathrobe wrap coat and those Victorian pearl earrings the color of brushed button mushroom caps, I'd surely wear them (with a ton of red lipstick) over the olive-green quilted satin pullover bomber-style sweatshirt and ash-gray silk cigarette pants and blue suede stilettos—to see my French lover who looks like Jeremy Irons. Wet autumn leaves on the New York City sidewalk, last night's rain running along gutters, *moi* scurrying along in the cold, shawl collar turned up, cheeks flushed, loosely piled-up hair juuust beginning to come undone like Meryl Streep's in *Manhattan* or *The French Lieutenant's Woman*. And Jeremy would embrace me in my coat that faintly smelled of him and of Coco perfume and he'd make my hair come undone . . . Afterward, I'd look like Carly Simon on the cover of *Boys in the Trees*—feminine, soft, romantic, suggestively sexy, languorously European, wearing a pale peach satin slip and pulling up a sheer silk stocking, and not like Carly on *Playing Possum*—open-mouthed and kneeling in a hard-core black lace teddy and black leather boots. My friend Amy, whom I met in Paris during my college junior year abroad, used to say, "Carly tries to sing like it's from her soul but it's really from her beaver."

Either way, what an orgasmic dream life can be!

CHAPTER TEN

The New Algebra

Our mutually bespectacled eyes met across the *huevos a la Malagueña*. He was cute-ugly in a Jewish, left-wing liberal, blond balding hippie, late Elvis sideburns, communist-pornographer kind of way. Let's call him by the initials T.P.—or Teepee—to keep it uncluttered. You'll see why. Trust me. Mami worked with him at St. Elizabeths. I was a fourteen-year-old ninth grader struggling—to put it mildly—with algebra. He was older and, true to his generation, wearing purple hip-hugger wide-wale cords on that Sunday afternoon in late May, a batik-y dashiki, and thick glasses with round gold metal frames as he introduced himself to me at brunch at our house and I introduced him to *huevos a la Malagueña*. Sizing him up, I knew his type instantly. Anything un-American was good, even Mami's half-assed huevos, of which Teepee enthusiastically ate two helpings. He probably loved—gag—Indian food, too.

"These *huevos* are just delicious," Teepee said, sitting next to me in the living room as I cleared away 893 stolen tchotchkes from the coffee table to make room for our plates and glasses.

There were about twenty other non-Cuban guests who appeared to agree with Teepee's inane *huevos* assessment. "I've had Spanish omelets before, of course. With potatoes. But these are really good."

"What you're eating *is* Spanish," I said, biting into a sesame bagel with cream cheese. Pretty soon I would have to go upstairs and sneak a cigarette. I'd begun smoking regularly. I liked True Blues, liked the recessed filter, liked the ritual of lighting a cigarette after meals and the relaxed-full way it made me feel. I loved it, actually.

"Oh!" Teepee said, smiling and nodding. He was Looking at me. Capital L. His eyes were blue but not true-blue. Behind him Carroll Sockwell's suicidal self-portraits in shades of blue gloomed in multiple heavy layers of dark oil.

"Spanish conquistadores and African slaves made Cuban food Cuban," I said, feeling *muy* authoritative and shaking stray sesame seeds off my lap. I was wearing my favorite sleeveless capri jumpsuit. Bright yellow with tiny blue birds and matching tiny blue bird buttons down the front. It was a bitch going to the bathroom, but I was still at the stage where suffering for style was a given, not to mention proof of my superior aesthetic sense.

"Wow," Teepee said. "Are you a chef? A historian?"

"I'm a Cuban," I said. "Cubans know their Cubanity. I just turned fourteen in December, what are you talking about? I just had my Bat Mitzvah last spring. I'm an official Jewess in high school, honey."

"Really? I would have never guessed."

"That I'm a Jewess? Why? You're a Jew, right? I'm a Jew, you're a Jew, everyone's a Jujube."

"No, no," he said. "I just meant you seem very . . . mature."

"Oh, I'm mature," I said, flattered by the observation—which I'd heard all my life—and by the fact that this man was actually paying attention to me. Growing up with parents with exceedingly low boredom thresholds, I've always expressed myself volubly—a lot more so than even a regular Cuban—and fast, because I've always felt the meter running out on parental attention spans. Really, my heart races and everything. (Or maybe it's just my metabolism; I usually think, move, and feel a lot faster than other people.) I realize not everybody is my parents. But I can't usually tell the difference when I'm on a roll or nervous because I have so much energy and so much I want to say. Excitement versus panic; sometimes I can't distinguish the two. Who can tell the dancer from the dance, right?

"Very mature, indeedy," I continued. "Don't let the zits fool ya. I have camouflaged them pretty well, though, I think. Which is really saying something since these are golf ballers."

"Golf ballers?" Teepee said.

"Pimples big as," I said. "I've got a friend at school who shoplifts makeup for me for cheap. She invited me to her birthday party at Kenwood Country Club? So my mother takes me and we drive around for three hours and never find it. We finally had to give up and go home. It's right there on River Road in Bethesda! Big sign and everything. Anyway, about the zits? My dermatologist Dr. Kanoff, she's terrifying. She's real strict and severe and she chain-smokes. I love her. The first time I went to see her and she examined me? We sit down in her office afterward and there's all this cigarette smoke everywhere. She gets out a prescription pad and on the top writes the word *NO* and underlines it, like, three times with a thick black fountain pen. Then under *NO* she writes this list: 'Iodine (shellfish, etc.). Chocolate. Nuts. Cola. Cheese. Whole milk. Fried anything.' I was, like, ahhh, just kill me now. I go, 'This is strict!' She goes, 'There are doctors who believe what you eat has nothing to do with how your skin acts. I'm not one of them.' So I decide to humor her to her face? But I go home and totally ignore this Nazi food list on the refrigerator. Mami always said to cut every corner. Like in *The Sound of Music,* 'Climb Ev'ry Mountain'? This would be her version: 'Cuuu-t eee-v'ry corrr-nerrr, searrr-ch high an' lowww . . .' So. I go back for my next appointment and Dr. Kanoff looks at my face and she goes, 'Are you eating any of the foods on that damn list?' I go, 'Well but we just got a brand-new Skippy. Crunchy. A whole JAR.' And she goes—she scared me half to death—'You'll either do what I recommend or you won't. If you won't, please don't bother coming back. It's a waste of our time. Good-bye!' I liked her ever since 'cause she sets rules and limits and she's not afraid to kick your ass when you're bad or lazy. I'm not used to that. My parents are out to lunch. They're in Europe and Israel and Latin America half the time. My father hates going anywhere, but my mother, well, you know how in the government they give you a lot of vacation? Plus she never gets sick so she's got shitloads of sick leave. *Joos eet or loos eet.*"

Teepee chuckled at my rendition of his coworker's argot. He said, "What I meant before was you come across as very precocious, very worldly."

"Well I do," I said. "The only thing I can't handle is algebra. It's my *nemesis.* Oh now there's a word I can expect to encounter somewhere in the verbal part of my SATs. I'm fourteen and I was supposed to be obsessed by my SATs ten years ago like my Sidwellemy classmates were, except I was busy fending off delusionals and paranoid schizophrenics at the time. So I'm way behind the curve. My mother's my nemesis, too, actually, but in a completely different way. Or maybe not in such a completely different way—"

"Are you wearing a citrus scent?" Teepee said in a strangely overfascinated tone.

" 'A citrus scent'?" I said, gently mocking what sounded to me like an arch, queer phrase right out of *Mademoiselle.* "No dear, I'm wearing Agua de Violetas. It's a traditional Cuban hair cologne. Violets and orange. 'Citrus.' Please."

"It's very alluring," Teepee said, leaning his face in my hair. "Mmm, I love it. It's wonderful. So fresh and feminine . . ."

"Just like using a turquoise bidet! That's what my mom does. Anyhow, I heard they give you, like, two hundred points just for getting your name right on the SATs. I know I'll fail the entire math section. I'm already failing Algebra I. I hate my school. I hate the people except my few friends. I don't have any friends who live around here 'cause I get shipped off every day in a yellow taxi with a morbidly obese trailer park woman named Tiny. The kids here go to public school in a normal yellow school bus. And the ones at Sidwell—that's my school, another nemesis—they don't live around here. Silver Spring is, like, beneath them. It may be beneath me, too, actually. I probably belong in New York City ultimately. What do you think?"

"Sidwell?" he said, reaching for his espresso. "That's a tough school to get into. You must be really smart. It's Quaker, right?"

"Yeah, right. Peace, love, tie-dye, and granola—and a big fuck you to you. They're all totally cutthroat phonies. Ivy League! Gotta go to the Ivy League or why keep on living! Gotta beat Muffy and Puffy and Huffy into the Ivy League! Muffy, Puffy, and Huffy— Snow White's little preppie dwarfs!"

Teepee laughed.

"Gotta live in Fat City and hit the club for cocktails and tennis and adultery after a hard week at the dwarf Ivy League law firm!" I continued. "Anyway. I'm pretty sure I'm gonna fail algebra. So tah-tah, Hahvahd Yahd."

"I was always pretty decent in math myself," Teepee said. "Mm, good espresso."

"Oh, my God, you're not gonna tell me you *like* it, are you? That it's . . . *fun?* Because I may throw up."

"Espresso?"

"Math."

"Okay, I won't say it's fun or I like it," Teepee said, touching my freckled arm. "Can I just say why it intrigues me?"

"No. I avoid discussing power raising, root extraction, and radicals whenever possible. Call me crazy."

"Because it's concrete problem solving. That's what life is, you know? It's why I'm a social worker."

Was it my imagination or had the fucker just moved in closer to me AGAIN?

I pulled away my arm and placed one of the 782 stray sofa cushions on my lap.

"Problems aren't always concrete," I said. "Maybe in algebra but not in life. Frenzy likes to say—this is such bullshit, how they describe algebra in their course curriculum, I love this—'Problems are related to those encountered in daily life.' Yeah, right. Anyway, that's why I love writing. Love love love it. People are the real puzzles, not equations. I think life is a mystery. My mother sucked at math and science because she was expected to, being a Latina and everything. Now we're in a whole different country, a *continent* instead of an island, but I'm still supposed to be just like her. I don't know which one to be loyal to, my mother and our little culture or my school and that whole world outside of *gringos.* 'Cause those Sidwell girls can all do algebra *and* ice skate. But I can write them under the table. So. I'm either algebraically incompetent or else really confused and frustrated. Which would make me *resistant.*"

" 'Resistant'? What do you know about resistance?"

"I *said* I was on the psycho kiddie ward—at your current institutional employer—when you were about the age I am now."

"I thought you were joking."

"Mental illness is no joke, dear. I almost got slashed and stabbed to death by a boy with a huge scissor on my thigh because I wouldn't kiss him."

"I don't blame him," Teepee said.

"Oh that's nice," I said dryly.

"No, I meant he had good taste," he said, tapping the second and third blue bird buttons on my jumpsuit for emphasis on the *good* and the *taste*. The blue birds were right between the 'zoomies. I should've put a cushion there, too. "Except I'd personally leave out the scissors. Staff or patient wanted you?"

"Patient," I said. "An older psychotic man of twelve. Only crazy boys like me."

"I'm sure that's not true."

"Well, another one broke my finger and practically killed my beautiful blue swan when I wouldn't kiss him. So sad, my swan. Papier-mâché. It got lost when we moved to this house. Typical. Long story."

"You inspire great passion in men."

"The blue swan killer was a black fourth grader named Maurice," I said, standing up. "Now he goes to Sidwell, too. So does my old Southwest D.C. friend Mara. It's weird. But most Sidwell boys don't like me. I repel the white preppie male element. They don't get me. Few straight XY chromosomes do. Starting with my father. My mother says he's not 'attun-ed' to me or my needs, not that she does anything about it or picks up the slack. The only female my dad can relate to is the wife unit. I have no relationship with my brothers. I don't understand their existence or its point. My little baby sister, Cecilia, is dead, in Cuba. I'm like Marilyn on *The Munsters. Voilà: Moi's* life in a coconut shell."

"Where are you off to?" Teepee asked.

I bent down and conspiratorially whispered, "To my room. To

smoke a True. To be bad behind my parents' back. I'm a kid, I'm in that adolescent angst phase. Hormones, you know. I rebel. I'm secretive. I'm pissed. It's what I do. It's my job."

"But your mother smokes. I thought you didn't want to be like her."

"That's the thing," I said. "Life is not concrete. I am *ambivalent* and *conflicted.* Kiddie ward words. *Adios,* dear."

"Can I come?" he whispered back, his face perilously close to mine.

"Shoe shine?" Eric said. "Just a dollar. Bargain."

My little capitalist seven-year-old brother had been making the rounds and now it was Teepee's shoes' turn. Eric had just saved my life. A grown man in my bedroom! In my boudoir! What would I have done? Was having seen *Klute* enough of a qualification to deal with a man nearly twice my age? And why was Teepee concentrating so much on me to the exclusion of the other guests his own age, anyway? It felt very odd, and I couldn't decide if it was odd cool or odd creepy. Excitement versus panic. I knew siccing Mami on Teepee would be fruitless. She'd *looohv* for him to go upstairs with me. After all, wasn't it Mami who'd rejoiced over the onset of my menses two years before because I could finally start dayteengh? As for Papi, he would always defer to Mami and, like her, would consider denying any guest in his home anything a social breach. If only Rebeca were here. That pygmy maid would slice off Teepee's moyeled pecker with one fell swoop of her cucumber knife and go right back to making *la ensalada* without missing a beat.

I heard Teepee ask Eric, "You know how to shine sneakers, buddy?"

I was already halfway up the stairs. I suspected our *intime* little tête-à-tête would make killer diary entry material, especially that shit with the buttons. *Good* and *taste.* My green Valerie typewriter awaited me. Me and my tiny but mighty right index fingertip.

I failed Algebra I.

Sidwell Frenzy said to get a tutor and I could retake the ninth grade final exam pass-fail at the end of the summer. Why the 'rents never asked Frenzy for qualified tutor recommendations is anyone's guess. Maybe Frenzy told them it was their responsibility, not the school's. Maybe Mami didn't feel like making the effort. Or maybe she didn't feel like coughing it up for one—it might have put a crimp in her personal retail budget. Hadn't she learned anything from the Tío Jaime Bat Mitzvah camera disaster, that you get what you pay for? Nooo, not as it applied to me. Enter Teepee, who offered to do it for *free*. (Well, almost. He wanted Cuban coffee in return for weeknightly tutorials.) *F-r-e-e*, Mami's favorite four-letter word after *f-o-h-k*. Did I have any say in this? Of course not, I was the *bruta* who'd yet again fohk-ed up. Did Mami perceive a possible conflict of interest, as she worked with Teepee, or the impropriety of having a decade-older man tutoring me alone? Of course not, he was a goo' frien' and he was *f-r-e-e*.

And so Teepee began pulling up on our cul-de-sac in his VW beetle just as Rebeca was serving the after-dinner espresso. We'd all sit around the round white mod table from Scan in the family room until the *café* was quaffed. The 'rents would go up to their bedroom to watch TV for the rest of the evening, and Eric and Big Red Al went downstairs to the basement to play or to their rooms to do whatever it was they did in there. Only Rebeca lingered, clearing the table, wiping it off, eating her dinner (as usual, after we did) at the breakfast counter outside the family room. She took a *real* long time eating whenever Teepee was around, and spent at least another hour dish-washing and leftover-storing and broad-spectrum kitchen anti-bacterializing.

"Why is she still here?" Teepee whispered.

Gee, maybe because she's the only one around with a clue?

"Who knows?" I said, batting my hand dismissively. "Who cares? She's a paranoid Ecuadorian. Ignore."

"She seems . . . hostile."

"She's never been the same since *This Is Tom Jones* went off the air. She goes around singing *'No es inusu-AL'* " all the time. She can't speak a word of English but she loves that Welshman. It's about, you know, the *feeling.* The universal language of luuuv."

"She goes around singing WHAT?"

" 'It's Not Unusual.' Tom Jones's theme song?"

"Whose?"

"You're kidding, right? 'What's New, Pussycat?' 'Delilah'? 'She's a Lady'?"

"Sorry," Teepee said, shaking his head. "At least you'll never fail popular culture."

"Pop," I said.

"Huh?"

"Not popular. Nobody says popular culture. Unless they're *un*popular."

"Hey, give me a kiss," Teepee whispered, leaning into me.

I grabbed my TaB—bet you didn't know it's officially kosher—and sipped through my straw. Too bad I couldn't smoke out in the open. I *so* needed a True. Who WAS this guy? Why was he saying this? *Hey, give me a kiss.* He was new in town but still, why wasn't he asking girls his own age to *Hey, give me a kiss?* What the fuck, should I go for it?

"On the cheek?" I asked.

"No," he said.

"Oh. All righty."

I heard Rebeca loudly clear her throat and slam down a pot.

"I've wanted to kiss you since the first time over the *huevos*," Teepee said.

"Well," I said, glancing at Rebeca, who stared back with fierce black crucifix eyes, "dream on—at the moment." It wasn't as if I'd never been kissed, but this was a Big Boy.

"Afterward," Teepee said. "We'll wait till she goes downstairs. And then you'll walk me to the car. It'll be our secret."

It was hard to concentrate exclusively on algebra after that. Teepee and I got into this flirtatious routine: He'd show up, drink the coffee, wait for everyone to disperse, and then punctuate the mind-numbing monotony of linear equations, inequalities in one and two unknowns, and quadratic equations with gradually deeper French kissing and touching. Then we'd walk to his car and kiss some more. It made me giddy. It was exciting and sexy and fun, and hell if it didn't liven up monomials and polynomials, not to mention what would have been an otherwise stupefyingly torpid, malaise-ridden summer in the Siberia of Silver Spring. Mami didn't *look* like a pimp, but hadn't she in fact brought us together? The Cuban in me said, *"Caramba!* Young brides are the best! Remember not the Maine, but the *tafetán color champán!"* The Jew said, "Oy! Remember Hitler at Munich! Then again, Teepee is Jewish and single!"

Teepee eventually complained to the 'rents that the family room was "too open and noisy and distracting for Gigi to absorb her lessons." Since both my parents lack that Cuban "Not with My Daughter!" gene, they immediately complied. Teepee and I began meeting downstairs in Papi's basement office. It was a tiny room but far from the madding crowd, and it had a door that locked. The office, however, was just across from Rebeca's bedroom. We could hear the humorless Andean warden pacing aggressively outside our closed door, clearing her throat and snorting and listening for prolonged silences. Rebeca could hover and eavesdrop in her overt, heavy-handed way, but she could not control everything. My parents may have told themselves and Rebeca that Rebeca was a member of the family, but in the end she was an employee, as I impertinently reminded her whenever in her pre-Columbian way she overstepped her bounds and tried disciplining me physically.

By summer's end, Teepee had me remedialized. On what I expected would be our last session, I walked him to the car to kiss him good-bye forever. Such sweet sorrow.

"Thanks," I told him. "It's been groovy. You really taught me some things, huh. 'The *new* algebra.' Haaa."

"You make it sound so final," he said.

"Isn't it?"

"Not right away." He was planning a move to a new job out West. "I still have some stuff at work to finish up."

"Oh."

After a pause, Teepee said, "I think we should be lovers."

"What?"

He had to be on junk. I mean, making out behind my parents' and Rebeca's backs was one thing; it was safe-naughty and something to do all summer long to break the ennui and I could feel as though I had a kind of boyfriend. But I knew I was no more ready for Actual Sex than I was for a Pythagorean Theorem.

"Yeah," Teepee said, holding my hands. "It's time."

"Don't you think I'm a tad YOUNG for that?" I asked, pulling my hands out of his.

But Teepee had listened to me too closely, become my confidant, and studied my mother at work and my family's dynamics at home for months. He was much too sophisticated and cunning to ever force himself on me the way you usually hear about these things. In other words, my mind had been the first organ he'd penetrated.

"I think," Teepee said, "you're every bit as much of a woman as your mother."

I somehow managed to pass my algebra final, and Teepee and I set up a "celebratory" tryst in Arlington, Virginia. I was a new tenth grader, at least a year younger than my classmates (thanks to Mami forcing me prematurely into kindergarten), with a big secret. The 'rents had taken off the entire month of September to vacation in Israel or Europe or South America. I really had nothing better to do; that was why I told Teepee yes. What I told *myself* was that we were like Romeo and Juliet, misunderstood by Society. Or like David and Bathsheba, sinful and secret. Or like Dimaggio and Marilyn, luminous December-May stars. Sure, I was jailbait and sure, Teepee was, like, about a decade older. But he'd marry me. He would. Probably. Right? *I'd* marry me. I was a saucy *señorita*, ready for love and ready to please my man, though I had no idea what that meant or what I was doing or why, exactly. I wanted to be like Bathsheba, bathing on the roof, and, to rearticulate Leonard Cohen, make my beauty and the moonlight overthrow him. But that was in another country. Virginia struck me as the kind of state that *would* take prisoners, take Teepee prisoner, I mean, if anybody found out. That would be bad, him being a felon. But look at the risk Teepee was willing to take, for *me*. Isn't that so touching? What had already been seduced in me from the neck up was *really* compelling. Was it panic or excitement?

As it always does for weeks after Labor Day in the swamp that is Washington, it was sweltering on the appointed day. I told Tiny and Rebeca I was going ice skating at an indoor rink after school with some girlfriends and would get my own ride home. It wasn't a total lie; I *had* taken ice-skating lessons the previous winter at the OUTdoor rink of a Virginia Marriott. My poor teacher. She had to practically hold me up the entire time. Meanwhile, all the other Sidwell girls whizzed by us like athletic sylphs, casting slush, although I will say my teacher remarked favorably on the bouquet of my Agua de Violetas. Nonexistent arches like mine cannot cope with laced-up, structured ice skates, we discovered. And no matter how cute your outfit is, you'll spend the entire time either grabbing on to the rail or your teacher, or on the ice on your color-coordinated ass.

It's a cold, hard country.

In Teepee's VW bug, as we crossed the Potomac into Dixie, I saw monuments and planes and the Pentagon and a lot of nondescript office buildings and high-rise apartment "complexes." Teepee and I were heading into Crystal City, a part of Arlington that's as antiseptically Republican as its name. I'd changed clothes in the bathroom after school, opting for a more event-appropriate cha-cha outfit. I borrowed my friend Mara's halter top with tiny red and white roses on it and paired it with my own very low-waisted hip-hugger jeans and caramel leather and wood platform sandals. I thought I looked like I had sex appeal, but this is what you think when you're fourteen. Teepee complimented me, and the radio played a Jimi Hendrix song: "You've got to be all mine, all mine . . . /ooh Foxy Lady."

Before going to Teepee's apartment, we stopped at a Safeway to scan cheeses and other snackies for a light postdefloration repast. It was over-air-conditioned in there, but my sweat glands were kicking into high gear *and* I had the shakes. Terrible combo platter. Teepee was asking me questions, something about "laughing cow" and "baby Brie." Bree? Bree Daniels? From *Klute?*

She has a cheese? God, I was so out of it—my body, that is—and floating up above us, looking down dispassionately. *This is weird, even for me. Only a man would think of food products at a time like this.*

What I recall about my First Time:

- The look of lust on The Prick's face (an expression I'd seen before only on Kevin, the scissor-wielding twelve-year-old St. Elizabeths patient: lidded eyes, open mouth—both kind of ick).

- How messy and sticky the white contraceptive foam was that The Prick had bought and how it mixed with my blood on my belly and thighs and the sheets (ick).

- How I could feel The Prick's but not my pleasure because he never pleased *me*, not that a Jubana who's told from day one that nice girls always wear panties—and occasionally bras—to bed would have any clue about her own sexual potential in the first place (ick).

- How after a shower I changed for The Prick back into my very Audrey Hepburn original school outfit (sleeveless tomato-red cotton linen pique dress with navy Nehru collar and vertical seams all across the bodice, and navy Ann Taylor ballet flats—both fabulous).

- How I passed on The Prick's *après-sexe fromage* product (ick), chain-smoked Trues (fabu), and chewed Dubble Bubble all the way home (fabu-fabu).

But here's the thing. You must have noticed I'm no longer disguising my feelings by referring ever so gently to Teepee. Let's call

T.P. The Prick. It's as close as I can get without terrorizing the lawyers to characterize the older man who deflowered me when I was below the statutory age of consent, The Prick.

The Prick dropped me off at the top of my long cul-de-sac and I walked down to my house, explaining to Rebeca that the school bus didn't do door-to-door. That night I stared at my face in the bathroom mirror. Did I look any different? Not really. My complexion still resembled Manuel Noriega's. Was I any good at Actual Sex? Well, I may have been clueless, but I must have been clueful enough to be asked back, because The Prick wanted to see me again. A Friday night sleep-over this time. Wow, I must be *pret*-ty hot stuff. *Foxy, foxy!*

It's the price of tafetán color champán, I kept thinking. The 'rents were still shoplifting their way through foreign countries, stealing fancy hotel ashtrays and sterling antique sugar bowls from assorted room services and restaurants. I told Rebeca I was spending the night at my Sidwellemy friend Pam Palmer's house. Pam's Chicago forebears built the Palmer House Hotel on East Monroe Street and the Cubs'—originally known as the Chicago White Stockings—first ballpark. (I went to the Palmer House with the flaxen-haired Pam once. It looked like a gilded castle inside. Almost twenty-five years later, in 1995, I stayed there again to do a cover story profile for *USA Weekend* on Heather Whitestone, the first deaf Miss America. It still looked like a gilded castle inside, only now they had modems, ATMs, and fitness centers.) Anyway, I told Pam I had this "friend" and he'd be picking me up *chez elle,* so she only had to entertain me for a little while, just a couple of hours until The Prick could leave work. Although people in our immediate circle didn't have "jobs," Pam didn't ask too many questions; she was good that way. She didn't come to Frenzy until ninth grade, in 1971, so despite her family's wealth and stature, she lacked that typical "lifer" conceit. But Pam was in the affluent cool white *gringa* girl gang, so I was cautious around her. Not that she was superior to me—please, the girl spelled *all right* as one word with one *l*—but still, hers could be a vicious little clique, so you never knew. (One time things got so bad between those girls and me that Mami Dearest stepped in for a "tehrahpooteek een-tehrvehnshohn cheet-chat." To their credit, *las gringitas* all came over on a Sunday morning, sat in the family room, and ate up all the *empanadas* and blintzes Rebeca had prepared ahead of time. *Then,* they proceeded to ignore me utterly and spent the rest of the time talking to Mami, praising her stunning red hair, her gorgeous figure, her adorable accent, her magnificent makeup, her dramatic ankle-length sleeveless cotton empire hostess dress with bright green, orange, hot pink, yellow, red, purple, and turquoise-blue birds of paradise, hibiscus, and ginger flowers all over it. I couldn't blame the girls for their adulation; I knew I couldn't compete with Mami. And academically, I couldn't compete with the Sidwellemies. I mean, I could, but even getting straight A's couldn't beat or even equal the reality of Eric's and Big Red Al's superior phalli. So why bother trying?)

The Prick picked me up and this time we went out on a proper date, to the Japan Inn on Wisconsin Avenue in Georgetown, the oldest Japanese restaurant in Washington. (I'd have preferred HH, but whatever.) We sat in the so-called teppanyaki room, where a severe chef cooked our food on a flat grill before us. I passed on the tako su (vinegared octopus and cucumber), and went with the miso soup, lobster, and filet mignon (The Prick did say to order whatever I wanted), and chocolate ice cream with mandarin oranges. The couple sitting across from us kept staring and whispering to each other. The Prick and I must have looked like a mismatch; he looked older than he was (balding head and newly sprouted Fu Manchu mustache) and I have always looked much younger than I am. My entire life, people, perfect strangers, ask me how old I am. Why is this? I'm guessing the combo of Jubana genes and tiny body with abnormally large head. Mami Dearest says, "Ees because joo look so johngh but soun' so ol'." So at fourteen I must've sounded sixty-seven but looked eleven— except with big bazoomies. *Foxy, foxy!*

In the morning in his Crystal City apartment I awoke before The Prick. I put on his cotton bathrobe and wandered out to the kitchen for coffee. A woman was sitting on a dining room chair, bent over and lacing up the long, thin white straps of her gladiator sandals. Her hair was a rat's nest from the back, and rolls of fat hung over the sides of her waist. On the table was an open pack of Virginia Slims Luxury Lights 120s, a lighter with peace symbols on it, a full ashtray, a key chain with 116 keys, a bad faux leather purse, and that morning's *Washington Post.*

"Hello?" I said.

She slowly turned with a haggard, puffy, pasty face that needed major concealer, loose powder, and blush. Not to mention mascara. Lipstick. Eyebrows.

"Morning," she groaned. "I'm Sylvia. Syl. And who're you?"

"Hey, you must be Gigi," said a tall, dark-haired guy with a beard. He was coming out of the kitchen with a pot of coffee in one hand and two mugs in the other. "I'm [The Prick's] roommate. Well, one of [The Prick's] roommates. There's three of us—sometimes. Coffee?"

"Sure," I said, completely perplexed. I knew The Prick had two male roommates, but who was the broad?

"Sylvia?" he said. "Coffee?"

"Yeah, coffee," she said, lighting a Virginia Slim. "Coffee and cash to go, honeybuns. Gotta hit the road."

"I'll be right back," he said, disappearing into a bedroom.

"How old are you?" Syl asked me. See what I'm saying? People never stop with the fucking question.

"Old enough, it seems," I said.

Syl gave a dubious look and went to the phone. I poured some coffee in the mug, lit one of her cigarettes while her back was turned, and walked over to the balcony window. Gray high-rises. A sliver of sky. Virginia sucked.

"Honey? Yeah. Mommy, who the hell ya think it is? Yeah. I'll be home in an hour, maybe less. Uh-huh. Baby okay? You watch him good? He did? Good. Anybody call? Huh? Watch that tone, missy. Yeah. See ya soon."

"So come 'ere and talk," Syl said. I sat down at the table and put out her appalling cigarette. Syl was right out of a Raymond Carver story. "Met him in a bar last night."

"Mm!"

"Yeah. But I think I had one too many. My head's crackin' open, I swear to Christ. You got any aspirin?"

"No, sorry."

"It's all right. That was my kid on the phone. My daughter. She watches my son while I'm workin'."

"Oh, uh-huh."

"You look about my daughter's age. What are you, like, twelve?"

"I'm almost fifteen," I said, wishing the Prick would wake the hell up.

"Well, 'cause if you was my daughter," Syl said, crossing her fleshy white veiny legs, "I'd beat your sweet little ass black-and-blue and twice on Sunday. You shouldn't be here. What're you doin' here, kid?"

"I don't . . . I don't really know," I said. "It's like a mini run-away, I guess."

"A what?"

"It's like running away but not all the way."

The roommate emerged. Syl wearily got to her feet, stuffed her cigarettes and lighter in her bag, yawned, grabbed her 116 keys, and talked to her "date" by the front door as I pretended to read the Style section.

"This ought to do it," I heard the man say.

"Cool," Sylvia replied. "Thanks. And hey, Gi-gi." She pronounced my name with a hard G, like *giggly*. She walked over to me and whispered, "Don't forget what I said. Black-and-blue and twice on Sunday. Get the hell outta here and don't come back. You don't belong. Don't be me."

They say the third time's the charm, but not for me, unless we mean "charm" ironically. The 'rents had returned from their holiday, and on this final, fatal school night I was late getting home from Crystal City, Vagina, as I'd taken to calling it. The Prick and I had fallen asleep. I'd called home around 8:30 or 9 P.M., assuring Mami Dearest I was fine and that the (fictional) ice-skating lesson had run long. She said, "Okay an' by de way I bought joo some cute panties on sale at Lor' an' Taylor." Panties. That's funny. But just between Jefferson Davis Highway and I-395, there was a major backup, and the drive that normally took half an hour was three times longer. So by the time I got home the frantic 'rents were in the carport, about to go to the police.

We went inside, Mami Dearest screaming and cursing Cubanly the whole way, Papi silent.

"WHERE WERE JOO?" Mami shouted. At the moment I was out of my body. And, now that I was alive, Mami could kill me. "WERE JOO WEETH [DE PREEK]?"

"Who do you think?" I said. Rebeca looked at me quickly

and scrammed. I knew it; she'd known all along what was happening and had ratted me out. AFTER I'd phoned home. What. A. Bitch.

"JOO WERE WEETH [DE PREEK]!" Mami yelled.

"Okay," I said. "If that's what you wanna think." My voice sounded mechanical, remote, unfeeling. Statue Voice. This would be bad. *Really* bad.

"JOO WERE! WHAT WERE JOO DOEENGH WEETH HEEM?"

"Doing?" I said, Bree Daniels irony mounting. "Talking. We were talking. I fell asleep on his couch. Sorry."

"¡LO PROVOCASTE!" You provoked him!

I sighed numbly and looked at my father. He was crying. He looked so sad, like a helpless, lost child. Sadder than the only other time I'd ever seen him cry, over a decade ago back at Las Casitas Verdes when the Bay of Pigs invasion failed.

"DON' JOO SEE JOOR FATHER EES CRYEENGH?" Mami shrieked. "DON' JOO THEENK JOO SHOULD GO AN' COMFORT HEEM?"

"Comfort HIM?" I said. That was rich. "No, I don't think he wants me to touch him right now."

"WELL, JOO ARE BEYON' GROUNDED!" Mami continued. "JOO ARE EEN SO MUCH TROUBLE! HOW DARE JOO WORRY PAPI AN' ME. HOW DARE JOO. JOO ARE SO SEHLFEESH. JOO LIE, JOO DEESREHGAHRD CURFEWS. JOO DON' CARE ABOU' ANYBODY BUT JOORSELF. JOO BLEW EET!"

"Oh *I* blew it," I said. My voice sounded tinny and flat, as if it were inside a cartoon bubble. "I see. Okay."

"NO!" Mami screamed. "JOO DON' SEE. EES NOT OKAY!"

She went to the kitchen, found one of Papi's ubiquitous prescription pads, and handed it and a black Magic Marker to him. On it he wrote the word *NO*, underlined several times. Underneath was a list:

TELÉFONO

COMPRAS [shopping]

CARRO [car]

AMIGOS

AFTER SCHOOL ACTIVITIES

FUN

Mami posted her NO prescription list next to Dr. Kanoff's. My parents couldn't even be original to punish me; they had to rip off other people's Nazi NO lists.

"READ DAT EVERY DAMN DAY," Mami bellowed, "AN' MEM-ORIZE EET. FOR LIFE."

My "life" sentence lasted a week, maybe two. The 'rents caved and we never talked about It again. It was as if It had never happened and The Prick as anything but a nice colleague and family friend had never existed.

Mami insisted on giving The Prick, whose move West was impending, a friendly farewell brunch. At our house. With all of us there. And the *huevos a la Malagueña.* It had a nice circular arc to it, I had to admit. We'd started out on a Sunday with the *huevos;* we'd end up on a Sunday with the *huevos.*

"If that *cabrón* comes in my house," Papi impotently told Mami, "I won't be there."

"We'll mees joo," Mami replied, exhaling her Kool and checking her manicure for chips.

I knew Papi would never NOT be there, but I could not believe The Prick would have the chutzpah to show up, especially after I'd warned him not to. While I was on parental probation, I'd talk to him from the pay phone at the Roy Rogers just up the street from Frenzy. He said he was sorry I got busted and that he loved me. He said we'd write to each other (my Temple Shalom Sunday school pal Sherry had agreed to have the Prick send his letters to her house). He said the world would never understand "our special secret love" and that we'd be together forever "someday soon." Even I had to laugh at that one.

"I think you'd better hightail it outta Dodge for good," I told The Prick. "It's not good for you to be around here."

But he came to brunch! Can you believe the balls on this Jew? Couldn't stay away from the *huevos,* I guess. I could barely look at him when he got out of his beetle, the VW bug that had transported me across state lines and the Potomac River to a changed self. By now I had turned fifteen and it was late May of 1973, May, that merry merry month. Peter Haldeman had left school the month before, and now this. All that remained of Juliet, Bathsheba, and Marilyn were bundles of illicit love letters sent to me in care of Sherry, which I kept hidden in a shoe box.

"Hey guys," The Prick said. My two little brothers were playing catch in the front yard. "How about we play after brunch?"

The Prick smiled at me. My stomach rolled, nauseating me.

"You coming in?" he asked.

I just shook my head.

The Prick shrugged and went in the house with the boys. I stayed in the yard and read my *Mademoiselle* under a cherry blossom tree. There was an article about taming your uncontrollable curls. Once I hit puberty my formerly *Funny Girl* Fanny Brice–straight locks hormonally morphed into a *The Way We Were* Katie Morosky Chia Pet. So I watered it, hoping it would magically remorph, maybe this time into Joni Mitchell or Cher hair. I blow-dried it on the hottest setting to within an inch of its life. I slept with it wrapped around empty coffee cans. I poured gallons of antifrizz conditioner on it. Not evenly vaguely like Joni or Cher hair. Still a Chia Pet.

I put down the magazine and lay back on the grass. The sun felt good on my face, slowing my breathing. I thought of my father

crumpling into tearful sobs and of my mother screaming how I'd
Provoked This. (Must've been my outfit. Who can resist big
bazoomies in small halter tops?) But the warmth of the sun relaxed me, bleaching away those immediate images. We really
needed a strong leader in my family who was a grown-up; my parents and I were having a terrible time raising ourselves. They were
two innocent little babies in adult bodies, aging children who
didn't know what the fuck they were doing or what to do or how to
help me or stop people from doing what they wanted to do that
might hurt us. Even now, Mami says, "Joo tol' me joo fell asleep
on hees sofa." When I ask what a fourteen-year-old girl might be
doing at 9:30 on a school night alone in an apartment in another
state with a twenty-something man, she says, "I don' know. What
I do know ees dat joo were a liar an' broke de curfews all de time.
Ehreek an' Alec never deed dat. Deyd come home early from der
curfews. Eef jood been honest I would have gone to de police an'
confronted heem. Are joo krehsee? I would never have heem een
dees house eef I knew!"

Hm. Teenagers discussing their sex lives with their parents?
Don't think so. Teenagers aren't even supposed to be *having*
sex lives. Sure, my Ivy-bound Sidwellemies were smoking pot,
snorting coke, drinking, and making out at parties to the
Stones' "Angie," Grand Funk Railroad's "I'm Your Captain," and
George Harrison's "Give Me Love (Give Me Peace on Earth)." And
who knows, maybe they were even sleeping together to Carole
King's soulful album *Tapestry*. (One girl in my class always carried
a diaphragm in her Coach purse. Her slogan: "I'm in demand. Better safe than sorry.") Maybe it's me, but I seriously doubt any of
those born-to-be-mild teenyboppers were off having sex after
school in Crystal City high-rise complexes with men past the legal
drinking and voting age who worked at mental asylums.

"See ya," The Prick said.

I squinted up at him. His head was nebulous and yellow. I got up, feeling dizzy. I knew he wouldn't. See me again, I mean. Juliet, Bathsheba, Marilyn, even *tafetán color champán*—all bullshit. Just myths, just stories and dreams.

"Good *huevos*," The Prick added, sounding and looking nothing like Jeremy Irons. "There's still some left."

The sun glinted off his dirty blond Fu Manchu mustache, the round gold frame of his eyeglasses, and the yellow car keys. He started to hug me but I pulled back, feeling sick to my stomach again. My mentor's mouth smiled. "And whatsoever Mouth he kissed—" as Emily Dickinson wrote, "Is as it had not been—"

Ick.

I couldn't believe it: For the first time in my voluble life I couldn't think of a single thing to say. It doesn't get much worse than that for a Jubana. I watched The Prick drive away, up the cul-de-sac, for the last time. I looked down. Sweat from my hand had puckered my rolled-up magazine. The American model's pretty face was obscured, as was part of the "Kill That Frizz!" headline.

All I saw was "Kill."

CHAPTER ELEVEN

Good Enough for Blanche DuBois

here's a Hemingway short story called "Soldier's Home," in which Krebs, a young college man from Oklahoma, enlists in the Marines in 1917, goes off to fight in the war on the Rhine, and returns to his mother's house two years later. The story is about the absurdity Krebs feels now, after what he has seen overseas. He's probably suffering from post-traumatic stress syndrome, though nobody knew from PTSS in Hemingway's time. Krebs's mother serves him breakfast: two fried eggs, bacon, and pancakes. As she tries to talk to him about getting a job and his plans for the future, Hemingway writes, "Krebs felt embarrassed and resentful as always . . . [he] looked at the bacon fat hardening on his plate."

That's how I felt in the weeks and months and years after The Prick, like a Hemingway character regarding the bacon. Which is why when Valerie, my beloved and beautiful WASP cousin, called up a few days after The Prick left for good and invited me to stay with her for a couple of weeks at the beginning of that summer, it was as if the clouds had parted and the sun came out.

Every morning Walter went off to work and Valerie made me what I considered the best breakfast in the world: hot chocolate, a bowl of chilled mandarin oranges, and a mini Sara Lee pecan coffee cake, heated just until the swirls of white icing softened. We'd spend the rest of the day at the Shoreham Hotel pool next door to Valerie's fabulous apartment, ordering lunch on Valerie's tab (which I doubt hotel owner and best friend Bernie Bralove ever made her pay), shopping at Lord & Taylor in Chevy Chase, and getting our nails done at their beauty salon. It was like being at some exclusive spa, the ideal therapy for a latter-day Jubana Krebs. If I just swam in my Valerie's Lord & Taylor boy-cut two-piece with black, white, violet, and aqua stripes; got sun; ate a club sandwich with my TaB poolside; had shiny coral-pink toenails; and didn't think about The Prick, I didn't get the queasy sensation. To me there was nothing worse. I could never understand bulimics. Anorexics, sure. But intentional vomiting? On your knees in front of the toilet with tears streaming down your cheeks? I knew a Sidwellemy who did that. It completely grossed me out.

The Braloves, Bernie and Alice, invited us to a dinner party at their house, the one Valerie had always told me was really nice. Valerie had bought me a new summer dress for the occasion, dark green with small white flowers, and new white leather T-strap Bernardo flats just like Valerie's sandals. Valerie wasn't kidding about the Braloves' *casa*—it was huge and gorgeous, with forest-green wrought-iron furniture in the beautifully landscaped large backyard, which was where we would be eating dinner. The family had two big, friendly, well-groomed dogs, golden retrievers, I think, that ran around with their big pink panting tongues and wagging tails. White-gloved waiters served cocktails (Shirley Temples for the kids) and lots of little tiny finger foods I could not identify on silver trays, and candles flickered and tiny electric white lights shimmered everywhere. It was like being a guest at Gatsby's, I imagined.

There were two long tables set up for dinner, one for the adults and another for the kids. If this were a typical Cuban party at the 'rents, I'd be seated with the adults, as I have always been. But here, I was seated with the kids, all of whom were preteen. The down cushion on my chair softly depressed under me. How magical. I'd never sat on down before. As waiters cleared the salad plates, I sipped my Shirley Temple and patted the dogs. The sun was setting, turning the already lovely garden into *A Midsummer Night's Dream.* I tore off pieces of dinner rolls and fed the dogs. Valerie had said tomorrow it would rain, so instead of swimming we could go see *The Exorcist* or *The Way We Were.* I overheard her and the other adults discussing the televised Senate Watergate Committee hearings, which everybody was obsessed with that summer. I didn't pay much attention; I just felt bad for Peter Haldeman and wondered what dinner would be, and dessert.

Something felt wet. I looked up. No rain. The evening sky was dimming but clear. I shifted on my down cushion and it was wetter. I pushed my heavy chair back, scraping its shapely verdigris legs across the gray flagstone. I pulled back my dress. In the shadowy light, black ink ran down my legs. I wasn't expecting my period. Actually, I hadn't had one for . . . was it two months? Three? I looked around. No one was looking at me. I got up and slowly walked backward toward the house. I'd noticed a powder room earlier, just off the patio's sliding glass doors. One of the dogs began sniffing the gooey path I was leaving across the stone tile, the other one began licking my legs. I broke away and hurried inside. My sandals squished and I tripped on a fluffy white double-pelt sheepskin area rug. Wine-red blood with things in it was everywhere, on everything—the white rug, me, and out past the sliding doors. It looked like a crime scene.

I locked the bathroom door behind me. More blood, heavier, thicker, and coagulated, oozed out of me, spilling out of my panties onto the floor, staining the (of course) white carpet. I carefully peeled off the panties and turned on the water in the sink. I soaked one of three very nicely monogrammed white linen hand towels in the stream and tried to clean myself. The towel turned red. I grabbed another and pushed it up between my legs. It, too, turned red. I sat on the toilet, and more big, soft, wet pieces and things plopped out, making the water splash up. I used all the toilet paper and flushed, but the wads of tissue clogged the tank. I tore off my ruined dress, the beautiful summer dress Valerie had just bought me that morning, and used it as a makeshift stopper. It was a terrible thing to do and see, but the tiny hand towels were obviously useless.

Jesus Christ, what *was* this?

I returned to the sink. The medicine cabinet mirrored a frightened ingenue in a slasher movie. Blood was smeared across my face, my glasses, my neck, my bra, my hair—how was that? I removed my glasses and washed my face but the bleeding below wouldn't stop. There was a shower stall on the far side of the room. I pulled off my bra and got in. *Great* water pressure. Of course it was. The Braloves are rich, they would only have a shower like this. Oh my God, so much blood on the imported white tile floor! It was just like in *Psycho* except a lot fancier.

What the fuck *was* this?

Oh, I just wanted to dissolve in that shower with the water and the blood and the purple pieces and red blobs and swirl right down that affluent drain. It was mortifying. I could never leave this bathroom! Well, hopefully I would die and then I wouldn't have to worry about it. The embarrassment. The clean-up. Who was going to clean this up? Who would? And how could I ever face the Braloves? I'd just ruined their beautiful rich people's house and my beautiful Bernardo Italian sandals. And, oh God, my beautiful Valerie. Oh God, Valerie. And Walter. Now the shower drain was clogged, too. Shit! This would be *bad*.

There was a soft knock on the door. Oh no.

"Gigi?" Valerie asked. Her voice was typically low, sturdy, and reassuring. "Love? It's Valerie. Are you okay?"

"Yeah," I lied, feeling weak. I wanted to disappear.

"Will you open the door?"

"No," I said, about to swoon. "It's not good in here."

"Just a crack? I'm alone."

"Really?" I could barely stand up.

"I'd never lie to you," Valerie said. "Come on, darling. It's all right, whatever it is."

"No," I said. "It's not." I may not have known what had happened to me, but the humiliation alone was enough to keep me in that bathroom forever.

"I'll help," Valerie said. "Please unlock the door, love."

I did. I saw the outer corner of Valerie's right eye. It was startlingly blue. I heard her gasp. I saw her blink.

"Hold on," Valerie said. She was taking off her sandals.

I backed up and Valerie walked in, closing the door behind her. We were both up to our toenails in my blood. It sloshed.

"You're going to be fine," Valerie said. "Just hang on while I—"

"I ruined my dress!" I cried. "And my sandals! My pedicure! I'm so sorry!"

"Fuck that," Valerie said, hugging me and imparting the perfume of her Joy. "We'll get all new ones."

I was wet from the shower and blood still flowed, though less, from the inside of my thighs. The embrace soiled Valerie's beautiful emerald green raw silk blouse, which she'd chicly tied in a knot at the waist. Though she'd rolled the bottoms of her white trousers, they too were red. Valerie reminded me of Jackie Kennedy in her bloodstained Chanel suit after poor dead JFK was killed.

"Hang on," Valerie said, cupping my pallid cheek in her tawny hand. Her smile was like her voice: gentle, strong, heartening. It always made you feel better, always less ashamed.

When Valerie returned, she'd cleaned up and changed into a sleeveless black cotton piqué dress, though she was still barefoot. I knew our hostess, the former ballerina Alice Bralove, was standing outside the door because I'd heard her and Valerie murmuring. Valerie held a pile of fresh folded bath sheets that smelled good, as if they'd just come out of the dryer. She dried me off. The blood was subsiding. An elegant pair of arms, dancers' arms, took the towels and handed Valerie two Kotex pads and an elastic belt with two hooks. I put it on. The arms handed Valerie a clean cotton nightgown. Valerie slipped it over my head. The arms handed Valerie a heavy cotton patchwork quilt. Valerie swaddled me in it like a papoose. She picked me up, my face curled into her chest, and carried me out to the car, where Walter awaited us.

"Ana? Gigi's not feeling well," Valerie said into the phone. I was sprawled on the Oguses' huge leather sofa, still wrapped in the quilt. Walter handed me my glasses—I'd forgotten all about them—and a cup of lemony sweet hot tea. "Oh no, she's going to be fine. Something didn't quite agree with her. You'd better come over. She'll be better off in her own bed tonight."

I could barely keep my eyes open.

"Your folks are on their way," Valerie said, taking the tea I was about to drop because I was so drowsy and drained. She lay down next to me and stroked short wispy tendrils off my forehead. I fell asleep to the comfort of Valerie's voice singing a familiar lullaby, our first Beatles song: "Oh yeah I'll tell you something I think you'll understand, / When I'll say that something, I wanna hold your hand . . ."

"Jesús Christ in the Ozarks!" the fiancé said when I told him the story.

This was the man I was going to *marry*. He was my best friend, the one I told everything to, entrusted everything to. I wanted him to know me, accept me fully, and love me even though (or maybe because) something really awful had happened to me a long time ago. It was part of me.

We were on my pretty ivory sofa, Paul sitting up and me lying down with my legs stretched out across his lap. I loved stretching my legs across his lap. I used to do it under the dinner table all the time. It relaxed me. When we were first dating, Paul sent me an E-mail saying, "I don't do feet [massages]." Well, maybe not the ex-rated wife's or any of my immediate (Paul's postdivorce) predecessors, but let's face it, I do possess cutlets. It's rare. So the fiancé got over that one.

"YOU SHOULD HAVE GONE TO A HOSPITAL IMMEDI-ATELY," Paul shouted. "PERIOD."

Paul loved using the word *period* at the end of his sentences and using it as a single-word sentence. In this eat-what-you-kill world it was the syntactical equivalent of walking softly and carrying a big swagger stick. Period.

"It wasn't Valerie's place to take me to the ER," I said, pushing my veal heel up against his Dino palm. "She would have never assumed that. Not her style to enlighten the 'rents or anybody else, for that matter. Women all around the world throughout history survive these things. Things are always coming out of us. The bleeding was stopping by the time we got back to her place."

"Geeg, you had a MISCARRIAGE from that fucking child RAPIST. You and Valerie were stepping on the PLACENTA."

I'd never said those words—*miscarriage, rapist, placenta*—to myself or to anyone else. At fifteen, I was still pretty immature and simply considered it a *really* freaky period. What wounded me more was having been fucked and forsaken. Sitting with Gramps fifteen years later, I talked about The Prick but I never mentioned the bloody aftermath because I still had a disconnect; I never associated having sex with what happened in the Braloves' bathroom. If I had, I wouldn't have consciously withheld it from Gramps, whom I trusted. But by the time Paul came along a decade later, something in me did make the connection, suddenly, on the sofa. Still, I was stunned when the "not observant at all" fiancé said those hard words out loud. I was already deeply in love with Paul. But this deepened my feelings and made me want to give him everything.

"Jesus Christ!" Paul continued. "You have to CONFRONT that sick twisted asshole pre-vert!"

"I never thought about it," I said, pulling away my pounded veal cutlets and sitting up. "Where would I even start? And anyway, maybe it *was* just a really really really bizarre period. You don't know and I don't know if it was really a miscarriage. Right? Who's to say?"

"Anybody can be found," Paul continued. He was on a roll. When he got like this I'd call it being on a Kaiser roll, which eventually got shortened to being on a Kaiser, which eventually got shortened to being on a Kais or just simply Kais-ing. "You have to do this. For YOURSELF. For your own CLOSURE."

"I don't know," I said, reaching for my trusty TaB and Parliaments. "It's not something you can 'close.' It's . . . internal."

"Seriously, all you need's a social security number or a driver's license."

"Are you gonna help me, if I do that?"

"What happened is OUTRAGEOUS," Paul said. "Period."

"Seeing The Prick won't change my life," I said. "It might make it worse. Will it undo the Braloves' bathroom? And everything else that brought us here to my Ikea sofa?"

"Yes," Paul said. "It will."

"Jubana goes vigilante on the trail of her baby rapist?"

"You're goddamn right."

"I think it would help more, or as much, if my parents—well, my mother, anyway—could say, 'We're sorry we let you down. We love you. This wasn't your fault. What can we do to help now?' "

"Good luck on that one," Paul said acerbically.

"I mean, I wouldn't have an answer but at least they would've tried."

"Well, *I'm* sorry," Paul said. "*I'm* sorry. It's not enough, I know. It sucks. But *I'm* sorry."

"Thanks," I said, kissing him. "That's really nice. I'm gonna go take a bath now. I feel dirty."

"Now? You're not dirty. Jubanas never are. They can never be. Jubanas always smell like Agua de Violetas."

The fiancé had a point, a sweet one. But I went and took a long, hot bath anyway. If it was good enough for Blanche DuBois it was good enough for me.

Period.

CHAPTER TWELVE

Size Matters

✧

*W*hy is your right sneaker torn?" Gramps asked as we
settled in for one of our sessions.

For this he gets $4 a minute?

"Ah, it's always torn there," I said, touching the thick white
cotton sock-encased ball of skin and bone protruding just beneath my big toe. It and its less protruding but still ballsy left foot
twin eventually break open all my shoes, not just the white
leather Keds I was wearing that day. That's partly why I have a lot
of shoes. You can share the love that way, spread it around.

"I know it's always torn there," Gramps said. "Why is that?"

"Just . . . normal," I said, shrugging. I flexed my short little
legs across the ottoman. They always feel better elevated.

"No," Gramps said. "It isn't."

"All my shoes tear there. My feet always hurt me. Since I was
little they have. In ballet class and ice-skating and birthdays. It's
'cause I don't have any arches. You know Reed Whittemore?"

"Who?"

"The poet Reed Whittemore? God, have you ever heard of anyone?"

"I've heard of fucked-up feet," Gramps said, regarding the PVCs. "And *those* are fucked up."

"Thanks."

"You're welcome."

"Reed Whittemore taught me Shakespeare at Maryland. He has this wonderful book of poems, *The Mother's Breast and the Father's House.*"

"*Oy.* She's starting in with the poetry. The Mamaleh's Booby and the—"

"Anyway. There's a wonderful poem in it—well, all the poems are wonderful—but there's this one called 'Writer and Reader': 'Let us agree . . . the truth is . . . in bare feet.' See? We wandered far off shore. You didn't think I could get us back but I did. So. Bare feet. He meant truth and freedom lie in bare feet. I've been squiggling out of my shoes and putting my legs up my whole life. It's what I do."

"Why?" Gramps said.

"'Cause I'm in pain," I said. "It relieves me. That ball there? It gets really red and sore by the end of the day. It vibrates, sort of like a pulse. The left one, too, but less. I guess 'cause the right side of my body is bigger and I use it harder."

"I want you to see a podiatrist immediately."

"You do?"

"*Oy!* Expansion! Thank you, God, the poem quoter heard me!"

To humor my expander, I went to see a colleague of Papi's. After examining, measuring, and reexamining the PVCs, the podiatrist said, "Your feet—the right one particularly—have developed swollen big-toe joints."

"Uh-huh," I said, simply happy to be out of my shoes and in a doctor's office. I've always loved medical environments. So clean and precise. Then again, I grew up on a mental ward, temped in Papi's office through my teens, and accompanied him on his rounds at the hospital.

"Bunions," the doctor said. "Permanent. The shape of your feet is deformed."

"No no no," I said, smiling. "They're pounded veal cutlets. They're *supposed* to be this way."

"What size shoe do you wear?" the doctor asked, typing something into his tiny laptop.

"Seven narrow," I proudly answered.

"You're a seven-and-a-half medium at the very least," he said. "Preferably an eight medium, depending on the shoe."

"Nooo. I'm a seven narrow. Tiny skinny little feet. I'm just like poor dead Jackie Bouvier Kennedy Onassis—so sad—and my mom."

"No. You're not. And incidentally, I always heard Jackie had big feet."

"That's impossible," I said, still smiling but getting hot under the 'zoomies, armpits, and face. I could feel the blood rising under my cheeks. If I'd been wearing my glasses instead of contacts they'd have been sliding down my nose. My heart raced. Fight or flight.

"You've been wearing the wrong size shoe for . . . twenty years," the doctor said, reviewing my patient form. "Your feet are malformed. Do you have pain?"

I suddenly thought of Rich's Shoes and Mami's former patient who worked there. I was attending a birthday party to seduce black parents into adopting me, and Mami bought me thin-soled maroon patent leather Mary Janes with squared-off toes and pearl button clasps. Beautiful shoes, painful shoes. *All de Jubanas mas finas* [most refined] *have dos same feet. Jackie Fohkeengh Kennedy has dos feet, okay? And her husband ees dead!* I flashed to "The Little Red Shoes," the folk tale I'd read as a newly arrived Jubana refugee child. How Karen had suffered for her own shoes! That girl had her feet chopped off by a hangman with an ax—while the red shoes were still on them! Maybe Karen was a *Chinita* (little Chinese girl) in extremis. Maybe her golden lotus blossom feet were, like, bound. Moral: You must suffer to be beautiful. Or, if you're already beautiful you must suffer, perhaps to become more humane, like Pilar and her mother in José Martí's poem "Los Zapatitos de Rosa."

"I don't really notice the pain," I told the doctor. "I mean, it's chronic so it's just part of me."

"Nordstrom has a good selection of widths," he said. "Go try on the right size. Don't wear narrow again. It's not for you."

For the second time in my life—and both times had happened under Gramps's tutelage—the top of my skull popped open. Metaphorically, of course. Then tears streamed down my face, not at all metaphorically. Great. The doctor regarded me with mystified sympathy.

"Can you fix them?" I said, looking around for something to blow my nose in. I really should carry those pocket packs of Kleenex like Papi. "Like, surgery?"

"Foot surgery is risky," the doctor said, handing me a tissue. "It can actually create more problems. No, no surgery. Just go to Nordstrom."

"But, you know, I just . . . I can't be that big. It's like monster feet. I already have an abnormally huge head."

"It's not that big," he said reassuringly.

I wasn't sure which extremity he meant. I dabbed my fallen, uncurled eyelashes. Shit! First it was tah-tah, Hahvahd Yahd. Now it was *adios, tafetán color champán.* Shit!

I went home and told Mami what happened. The chic denial freak couldn't argue with a *doctor,* a board certified, highly experienced *podiatrist* and *surgeon*. She exhaled her Kool smoke impatiently, raised the volume on *Now, Voyager*, and returned her concentration to applying Revlon top coat to her long ice-white talons.

"So?" I said, a little edge in my voice. Hey, I'd earned the right. I officially had eternally fucked-up feet, thank you very much. It was like being blind before I got glasses; I assumed everybody danced into cars after seeing *West Side Story* or *My Fair Lady* and that their feet hurt while they did it. "What do you think of THAT?" I continued. "Turns out I've been wearing the wrong size shoes since I was—"

"OKAY, DAS ENOUGH!" Mami exploded, momentarily lifting the wet wand's tip from the top of her nineteen-inch-long left thumbnail. God forbid a smudge. "OKAY?"

I grabbed the remote off the arm of her rocking chair and turned down Bette, who was in the process of having a fabulous fight with Gladys Cooper, her mother. It was a shame, too, because I love Bette Davis in that movie, even if Pauline Kael did think Charlotte Vale, Bette's character, was masochistic. Okay, maybe she was, but so what? Bette's clothes were great after her ugly duckling–to–swan metamorphosis, inspired by an expander and later on by a married man, especially that scandalous plunging black number she wore with a fresh gardenia to host her first postmetamorphosis dinner party. Not to mention everybody smoked constantly.

"Mom. What is your problem? The doctor YOUR HUSBAND recommended SAID my poor feet—"

"I don' HAVE a problem," Mami said. "OKAY? What I resent ees joor sahrkahsteek TONE an' de BADGEREENGH. Because I DON' DESERVE EET. I theenk JOO save up all JOOR bad feeleenghs johs for ME. An' I don't LIKE eet."

"Okay, then cough up the Nordstrom card," I said.

"*¿El que QUÉ?*" Say WHAT?

"Expunge your guilt. Dollars for pain. Tit for tat. Quid pro quo. Eye for an eye. You get the gist."

"No, I won'!" Mami said.

"You 'won'? You won what? You didn't win anything."

"I SAID I WON'T. T-T-T."

Wow. English as a second language self-correction. First time in . . . ever. The woman who refused to say camp and kind all the way because life ees too damn chort for dos final consonan's? This was progress. Well, lingually.

"In that case I'll just hobble back to the *arroz* paddy on my hands and knees like a crippled *Cheenah* from the tenth century," I told her. "It's right up there with clitorectomies and bride burning. Advanced."

"Das Japan," Mami said, turning Bette back up. Bette was saying, in a wonderful moment of abrupt self-discovery, "I'm not afraid, Mother, I'm not afraid!" I love that line. It's like in *The Moor's Last Sigh*, when Salman Rushdie writes, "I'll tell you a secret about fear: it's an absolutist . . . Either, like any bullying tyrant, it rules your life with a stupid, blinding omnipotence, or else you overthrow it, and its power vanishes in a puff of smoke . . . I stopped being afraid because, if my time on earth was limited, I didn't have seconds to spare for funk . . . *I must live until I die.*"

"No," I told Mami. "That's not Japan. That's China. Geishas are Japanese. They don't have those bound feet." (Pounded feet maybe, but not bound.)

"But de 'rectomies?" Mami said. "Ahfreecahn. Dehfeeneetly. An' dos fire brides, deyr een Eendeeah weeth de dowries. Well, weethout dem. Das, like, Feefth World mehntahleetees. Now get out der an' make eet a good damn day! Lohv joo more!"

CHAPTER THIRTEEN

Letters from Madame Emma B. Ovary

*M*y sex letters to Woody Allen, which I foolishly hoped would lead me to the *tafetán color champán*, were always Jubanically *caliente*. But it was the envelopes that got him. They were killers. They took hours and hours to do, these intricate interconnecting, overlapping, intersecting collages of glued-on curvaceous thighs in black fishnets from Victoria's Secret catalogues, dried and silk flower petals plucked from Mami's infinite array, and fragments of poems and fiction from the *New Yorker* (unless it was irresistible copy, I avoided using newsprint; doesn't hold up). These inspired *chef-d'oeuvres* were my insignia, so Woody—or as I called him, my leetl Voody—would know it was from La Gigi—or as I called myself to him, Madame Emma B. Ovary. The correspondence from Voody to *moi* was never decorated in kind. But then, he's Voody. Voody doesn't have to do these things. He just used his courier new font letterhead brown-paper-bag-like stationery and dashed back sexy, funny, flirtatious notes in his thick black pen, often punctuating his bouncy prose the Emily Dickinson way, with more dashes than periods between the lines. Exchanging lust letters with Woody Allen was much better than phone sex. Less cheese, more filling.

But I'm getting ahead of myself, *n'est-ce pas?* You're wondering exactly *how* did Woody Allen, my prospective husband, and I hook up? Well, the *New Yorker.* No wait, Beaver. Yep. Beaver. Beaver conjoined *nosotros.* It's romantic. So contain your gonads and I'll explain.

Everett Lloyd Kayhart (B.S., United States Merchant Marine Academy; B.S.F.S., Georgetown University) was hacking. *Hack-hackhack.* He was coughing his way through my Frenzy academic career and my life. The ever supportive chain-smoker Sidwell Frenzy guidance counselor was shaking his head as he looked at my SAT scores and my GPA. He exhaled a manly-man billow of smoke in my eleventh grade face.

"You'd be lucky to *(hackhackhack)* get into Maryland," he said gruffly, interspersing every few words with his customary hack. It was like Berkeley Breathed's Bill the Cat upchucking a hairball. "What a waste. Think your parents sent you *(hackhackhack)* here so you could wind up at a goddamn public school?"

Since we both knew this was a rhetorical question, Kayhart just shook his head, exhaled disdainfully, and *hackhackhacked,* this time into a monogrammed handkerchief. Was there anything these goyim didn't monogram? It was terrifying. The real question was how I would save myself from this hacking, snorting, merchant marine, Marlboro man, murderous *gringo* bitch who never knew windows opened. His teeth and his nails and every other surface were covered with a stain the color of curry and cured tobacco.

"At this point," Kayhart said, now snorting instead of hacking, "I'd urge Beaver."

"You'd *what?*"

"College, dammit. Beaver *(hackhackhaaack)* College."

Hello, did I *look* like a rodent? First I'm a *gusana* for leaving Cuba. Then I'm a tropical termite for being Cuban. Now I'm a semiaquatic herbivorous pudenda for sucking at Frenzy? Nice.

"Beaver?" I asked. "What is that, like, Beaver A&M? Agricultural and Mechanical?"

"What?" Kayhart sputtered, shrouding the EKL monogram on his handkerchief with a blob of pea-green phlegm. "What the hell are you talking about?"

"Or Beaver Ancient and Modern?"

"Anders, are you HIGH?" he yelled. The effort triggered a multiple hack attack.

"Would that I were," I said with a sigh.

"You're applying to Beaver early admission," Kayhart said, wiping the side of his thin mouth. "Risk reduction. Containment."

I'd been looking at smallish private liberal arts schools like Swarthmore, Barnard, Georgetown. When I told this to Kayhart he couldn't stop guffawing. He wiped his soggy eyes with his now soggy handkerchief.

"Oh man, you're killing me," Kayhart said, reaching for a fresh Marlboro.

"Would that I were," I muttered.

"You go look at Beaver. And you keep *(hackhacksnooort)* looking at it until it lets you in. Beaver's the only private college in the Northern Hemisphere that'll accept you."

In other words, I was Beaver bound. At the student bookstore they sold T-shirts that said "Save a Tree / Eat a Beaver." I really should have sent one to Kayhart to express my gratitude. Located in Glenside, Pennsylvania, a Philadelphia suburb, Beaver had, by my 1975 matriculation, just recently gone coed. (Non-Beaver people would ask, "Beaver? What is that, an all-girls school? Haaa.")

By 2001, however, too many Internet porn hits made the school change its name to bland-but-safe Arcadia University, decades too late to save me from the obvious snickers.

It was the spring of 1977 and we'd just read Flaubert's *Madame Bovary* in my Beaver College European lit class. I was completely infatuated with the novel. To that end, my professor brought in a recent issue of the *New Yorker.*

"Check it out," she told me. "There's a short story in there by Woody Allen that you'll appreciate."

It was called "The Kugelmass Episode," in which a Jewish New York City professor goes back in a literary time machine and has an (ultimately disillusioning) affair with Emma Bovary, his favorite fictional heroine. Everybody's heard of Woody Allen, but I never knew he could write so well. I flipped over the story.

When the semester ended, I went home to prepare for a semester in New York University's program in Paree. An old Sidwellemmy friend, Nicole, suggested we take in Woody's *Annie Hall,* which was just out. I fell in love. Diane Keaton was so funny and fresh and Woody was so smart and endearingly neurotic. I decided I had to reach that crazy Jew who could, it seemed, do anything well. And when you're a teen you think you can do anything, too. Inexperience informs impertinence.

I took out the Crane's, the heavy 100 percent cotton pale blue filigreed stationery, and my Waterman fountain pen, and began writing: "My leetl Voody, Yonville [where Madame Bovary lived as a married woman] sucks weethout you . . ." I told him how much I missed him, missed his keesees, his "fohnny leetl *je ne sais quoi*-ness." I told him I'd had a change of heart, that I wanted him back. I signed it, "Love and amour, Emma B. Ovary."

I attacked the matching envelope with my quintessential collage, leaving room only for stamps and my and Woody's addresses, and collapsed. The next day, a family friend called. He happened to know someone at United Artists in New York and suggested I send my letter up there and request it be forwarded. I thought, what the hell, and I mailed that sohkehr.

A week passed, and on a sunny Saturday afternoon, I was just getting in, my arms full of drugstore shopping bags stuffed with items for my trip to Paris. I remember I was pulling out a plastic bag of cotton puffs when the phone rang. The voice on the line was unmistakable. I dropped my puffs.

Voody and I talked for a long time, possibly two hours, possibly fifteen minutes. Impossible to know. He was sweet, friendly, warm. He told me he usually didn't go through his fan mail himself, and that his secretary brought him my letter because the envelope caught her eye. He said he found my letter original and witty, and he said a lot of other very kind things. He wanted to be my correspondent. Mine! So I gave him my address in Paris at the Cité Universitaire, a complex for international students in the 14th arrondissement. Woody gave me his home address on Fifth Avenue and implored me to "make all your letters as hairy as this one."

" 'Hairy'?" I asked. I had no idea what he meant.

"Yeah, tell me about your first orgasm in Paris," he said.

"Um, okay."

When we hung up I screamed to the 'rents what had just happened.

"Were joo JOO weeth heem?" Mami asked. "Because joo have to be joo."

"I don't think Woody Allen would make a very good son-in-law," Papi remarked.

And so over the summer (and the following couple of years) I sent my leetl Voody postcards, letters, French versions of Woody

Woodpecker comic books, and yes, I informed him about my first orgasm in Paris. I described the room, the ambiance, the weather, and closed with this: "Would that there had been someone else there to enjoy it with me."

To my Parisian and domestic friends, I became a kind of unfamous celebrity. My delight over Voody was infectious, and I cherished what he and I had so intimately created. A fantasy. My fantasy was that Woody Allen and I were *affiancé*. His fantasy was . . . well, obviously, I don't know what his fantasy was. In all, I wrote dozens of times, sometimes as *moi*self, but usually in the persona of "Emma," because as Emma I was much less restrained. (When the Soon-Yi thing hit the fan back in 1992, I wrote about Voody's and my romance for the *Washington Post*. It was called "Woody and Me: A Love Story. Sort Of.") One of Voody's letters included a casual suggestion that I drop by and see him the next time I was in the city. Mental note to self: *If ever you leave Paris, you'll always have Voody.*

North American reentry *après* France was tough. *How're you gonna keep 'em down on the farm after they've seen Paree?* Hey, how're you gonna keep 'em down on the farm after they've seen the FARM? I wrapped up my final English and art history credits at the University of Maryland in College Park (which, after Paris, was like being in Bombay), and emerged in 1980 with a bachelor's degree and scant possibilities for employment. I hadn't taken a single class in journalism, though I had written a regular column for the *Beaver News* called "Anders Ganders." And while my regalia may have been a pair of webbed feet, it was a start. After a brief waitressing apprenticeship at Angelo's Wayside Inn in Silver Spring, Maryland, I decided to go work at the *Washington Post*. I could probably write some stuff there. It never once occurred to me that most people didn't just one day decide to go do that and then go do that.

So I did that.

A year later, I went to New York to visit Mary Lou, my platonic Frenzy boyfriend Rob's older sister. She knew the whole Woody story, and on a cold, cold Manhattan night, we impulsively decided to go see him in the flesh at Michael's Pub. Woody used to play jazz there on Monday nights. The club, on East Fifty-fifth, wasn't far from Mary Lou's place. While we were walking the few blocks, Mary Lou put on a pair of sunglasses. To me she was the epitome of common sense and cool, so I put my shades on, too. The city looked dark.

"Why are we wearing our sunglasses again, Lou?"

"Because the wind hurts our eyes," she replied.

Michael's Pub made my heart pound with apprehension. I was abruptly ambivalent about this whole thing. Laughing girls and the clatter of dishes and Woody there with his clarinet and his hair looking so red under the hot lights—it all made me feel like a Jubanique interloper.

It was suddenly too real.

After the set, I shakily got in line to greet Woody, as did almost everyone there. From her chair, Mary Lou signaled a thumbs-up to me and smiled encouragingly. I looked back at her as though it was the end of something. I noticed a pale, skinny brunette sitting next to Woody, smoking a cigarette and aloofly sipping Perrier. She looked about fifteen. I smoothed down my long nubby gray-and-white tweed wraparound sweater. It was my turn and there he was, glancing up at me from behind his imposing, silly glasses. I felt nauseated.

"Hi," Woody said. "Who should I make this out to?"

"No, nobody," I said. "Nobody. This isn't . . . I don't want . . . not an autograph."

"Okay," he said, composed.

"Wait," I said. I felt myself sinking. I took a collaged envelope I had prepared out of my cardigan pocket and proffered it. I would have recognized one of Voody's envelopes instantly. Surely Voody would recognize one of mine.

But Woody Allen just stared at it, blankly. His pubescent date exhaled languidly and looked away, bored out of her mind.

"It's from Emma," I said. "Emma B. Ovary?"

"Emma," he said. It wasn't a question. He looked baffled.

"Yeah," I whispered to the floor. "Emma. You know."

He acted like he didn't know. The man in line behind me was snorting his impatience. The skin under my 'zoomies was soaked, as were my armpits and my face. No amount of concealer or powder, not even industrial strength, could've quelled it.

"What am I supposed to do?" Woody asked me, holding the moist envelope.

"I don't know!" I cried. "'Bye!"

I pulled away, full of shame and tears in that blurry room. As Mami Dearest would've said, "Joo blew eet." Yes, I had tendered enough information for a man who'd been nursing a transcontinental love affair with a woman he found fascinating. But probably not for a celebrity who'd scribbled a few playful notes to a besotted college student.

Was Woody being coy? Was he embarrassed I'd shown up unannounced while he was out on a date? Did he even remember who I was? Was it that he just didn't like my sweater? My curvaceous Cuban ass? My abnormally large head?

"I want to go home now, Lou. Okay? Please?"

Mary Lou understood. We put our coats and scarves and gloves on.

"Did you tell him who you were?" she asked.

"Yes," I said. "Yes."

"Weird. Oh well. You tried."

"I know," I told her. "I don't know what happened. I felt closer to him when I didn't meet him. Woody'll never be my husband now."

"Thank God," Lou said.

When we stepped into the street, the wind had picked up. It was biting. Mary Lou put her shades back on. I started to also, then didn't. I was still crying. Was it Woody or the wind? Whatever it was, it was good to be walking away from it.

CHAPTER FOURTEEN

Little Blue Box

For a Jubana who's never fantasized about her wedding day, it was amazing how effortlessly I clicked into my alter ego, Lulamae–Holly Golightly. Being in love and *Breakfast at Tiffany's* had brought me here, to Tiffany & Co. on Wisconsin Avenue in Chevy Chase, to check out engagement rings with Paul. Yes, my prehistoric beloved Dino boyfriend had become my prehistoric beloved Dino fiancé. Fee-*AAAHn*-ceh. Such a great word! Boyfriend can't compare to fiancé. It's so much more . . . *je ne sais quoi.* Wahndehr Brayt as opposed to *croissant.* Wal-Mart as opposed to *Tarjay.* Schlitz as opposed to *Champagne*—any *Champagne.* Or maybe I just like saying French words a lot.

So across the glass counter we leaned, and I knew we would lean that way forever. It was so beautiful and we had rain in our hair and I couldn't remember a happier moment than on this Saturday afternoon in October, with me in my single-breasted black patent leather raincoat with the electric-blue wool lining, and Paul in his lined khaki trench coat. The sky'd been Jeremy Irons moody since morning, when we'd drunk our *cafés con leche* and chewed our toasted bagels—sesame for me, pumpernickel for *lui*—with cream cheese, sitting at the same family room table where Mami Dearest and I'd fought over the wedding guest *lees* and reception appetizer selections just weeks before. (The prospect of matrimony may be the only time in Juban life when people will actually, like, *plan ahead*.)

Lucida Lust. Who knew I had it? Must've been a *very* latent gene. Because when I laid my Helen Keller eyeballs on Tiffany's brilliantly blinding Lucida ring, I could suddenly, miraculously SEE! I was cured! *Tafetán color champán*—bring it on! A-a-a-mazing grace . . . Really. It was sick how much I loved that ring. First of all, the name. It's Spanish. Well, it's actually Latin. Close enough. It means bright or shining or clear. *Lucid*, get it? LOO-see-dah. The distinguished, continental man helping us said it meant "the brightest star in the constellation," but *en Español* that would be *lucidísima.* Whatever. Lucida was the One.

"Le plus lucide," I told Paul, who's sort of multilingual. He took off his raincoat. Perhaps I sounded a tad too enthusiastic; it seemed to be making the reptile fiancé sweat. *Do* reptiles sweat?

"Oui," said Paul.

I tried on a one-carat Lucida engagement ring in a platinum setting with a matching diamond-less Lucida wedding band, also in platinum. (I knew the 18-karat yellow gold setting would flatter my skin more, but the hell with that. Besides being the most durable of all the precious metals, platinum is, well, platinum.) The rings were truly, flawlessly, simply orgasmic. The closest mere mortals can get to this *mise en scène* is at www.tiffany.com/expertise/diamond/rings/combination_lucida_ring .asp?ring=dia band&. What was weird about it in real life was that nothing was weird about it. Because it was *Tiffany.* Because it was *Paul.* Because it was *right.* After all, everybody knows nothing bad could ever happen to you with a Dino-man in Tiffany. Plus, the man who loved Holly Golightly for the Lulamae she really was, in her soul, was named *Paul.* Plus, Holly realizes Paul is the One while they're *standing in the rain, holding each other and wearing raincoats!*

"This is beautiful!" I cried. I put my hand in the fiancé's paw so he could have a closer look.

"Nice," he said, nodding. Men know nothing. They just want to do this so they can move on to the next thing, which in Paul's current context meant either shrimp fajitas or *salmon a la parrilla* with shrimp and marisco sauce. We'd made plans to meet up afterward with my old *Washington Post* friend Joe McLellan at Cactus Cantina, our fave Mexican restaurant in D.C., just north of the National Cathedral. See? A very wedding-y weekend: Tomorrow morning we'd have our first appointment with Rabbi Bruce in his study at Temple Shalom for premarital counseling.

"We put as much emphasis on design as on the quality of stones," said Tiffany Man. He reminded me of David Niven, suave and cosmopolitan with his thin little mustache, and John McGiver, with his balding head and vaguely superior manner. "The Lucida provides a marvelous alternative to a strictly traditional or modern engagement ring. It's a lovely blend."

"It IS!" I cried. I glanced at Paul. Face shiny but body still vertical. Good.

"This diamond shape is called the Lucida cut," Mr. Niven-McGiver continued. "It's our own creation. A mixed cut, not quite square, rather like a round brilliant on the bottom, so it has a lot of sparkle."

"It DOES!" I cried. I sure was crying a lot in there. Hey, my Lucida was worth crying over. It was *lucidísima.*

"The wide band gives the ring security," Mr. Niven-McGiver

expertly explained. "It's a more contemporary style. Very sleek. Graceful. We're known for a few specific designs. This one was introduced in 1999."

"You're not gonna stop making it, are you?" I asked, slightly alarmed. "It's like a perfect little sculpture from MoMA, but it's wearable!"

"We'll always carry it," he answered, discreetly slipping me his card. Jesus, even the card stock was thick and creamy and gorgeous. Of course it was. "It's part of our permanent collection."

"How are you doing?" I asked Paul. "Are you okay over there?"

"Sure," Paul said.

"Really?"

"Why?" Paul asked.

"Could you give us a moment?" I asked Mr. Niven-McGiver, who nodded elegantly and glided to the end of the counter. "You. Divorc-ed Dino."

"Sí?"

"You. You are a . . . um, how shall I put this . . ."

"Cheap bastard?"

"Right. Exactly. Thank you. So why aren't you, like, passing out with fever or throwing up or running away screaming?"

The fiancé took my tiny fiancée hand in his huge big one and kissed my palm, careful not to mar MY LUCIDA.

"I love you," he said. "And yeah, it's expensive."

"Dear, it's, like, five digits, this combo platter."

"It's Tiffany," Paul said with a shrug. "And it's what you really want, isn't it?"

"Yeah," I admitted. "It is."

"Okay, then. This is, I hope, a once-in-a-lifetime purchase. That's what savings accounts are for. You know, an investment. You'll have it forever. I want you to have what you want. I live to serve, isn't that what you always tell me my place is? My raison d'être?"

"It's becoming part of my haaand," I said with a swoony sigh, clutching my hands to my chest like a nosegay. Lucida had to be better than crack.

"You know, Martha never liked diamonds," Paul said, trying on one of the two plain platinum Lucida wedding bands Mr. Niven-McGiver had left on the counter for him. The plain ones are unisex. Men usually choose the 4.5 millimeters width, or the 6. Both were fabulous on Paul and the most expensive sohkehr was only $1,000! So cheap! That's like one little magazine story for me.

"The ex-rated wife didn't like diamonds?" I said, incredulous.

"Nope. Not her style. We had our gold bands made. We designed them ourselves."

"You are so lucky to be with me," I said, closing my eyes and shaking my head.

Paul kissed me.

"Hey, is this really okay?" I asked him. We were, after all, shlubby journalist Democrats. "Are you sure you can afford this? Because if not, we can look on Sansom Street. Isn't that where you told me in Philadelphia that's, like, a whole city block, and all the members of our tribe sell the pet rocks there?"

"Aw, come on," Paul said, slipping off his Lucida. His naked hand looked sad without it. Just your basic reptilian paw. "Every woman wants that little blue box experience."

"Can't put the rain back in the sky / Once it falls down / Please don't cry. . . ."

"Ooo," Paul remarked. "That's good. 'Can't put the rain back in the sky.' "

We were sitting in my car in the Giant Foods parking lot behind Cactus Cantina, listening to Lucinda Williams's CD *Essence*. This song was called "Are You Down?". We'd arrived early and were killing time waiting for Joe to join us. I'd loved Lucinda for years, especially *Car Wheels on a Gravel Road*. I loved introducing the fiancé to the music I loved; musically he was kind of a one-trick pony. For him it was jazz, jazz, jazz.

"Lucida's amazing," I agreed, exhaling my Parliament and sucking my lime Tic Tacs.

"Who?"

"Lucinda Williams."

"You said Lucida."

Well. Now that we clearly knew our ring's names—and sizes— Paul and I would buy each other's separately and act surprised when we proffered our little blue boxes.

"Can't force the river upstream / When it goes south / Know what I mean?"

In Rabbi Bruce's cozy study, I was excited and loose and giddy. Paul was grim and vigilant and tense—arms gripping his chest and legs crossed for the entire two hours. He appeared tortured.

"This isn't the Inquisition," I whispered to him as Bruce gathered some papers. "You're allowed to take off your raincoat."

"I'm fine," Paul said, shifting mournfully in his chair as if it were electric and he'd spent the last decade on death row and all appeals for clemency had been exhausted.

Bruce talked with us about the history of Jewish weddings— Reform Judaism is about informed choice, and no limits were put on what was possible.

"You don't plan a wedding and make decisions as if the last

276

four thousand years of Jewish weddings didn't exist," Bruce said. "So we'll consider that carefully. At the same time, you should be free to determine the most meaningful content. That's liberating: It creates awe and respect."

"In with the old, in with the new," I said. "I love it. Individuality! Right, sweets?"

Paul looked stricken. I attributed this to the fact that my Jewish fiancé had never had a Jewish wedding. Martha the ex was a WASP who'd converted to Catholicism.

"My primary point," Bruce said, "is that only the bride and groom know what's truly meaningful to them. That's what I'm after. The only thing we always include in the ceremony is for each of you to tell the other, in Hebrew and in English, 'Be consecrated to me with this ring, according to the law of Moses and Israel, according to the law of God.' "

We moved on to the history of our romantic relationship. Bruce wanted to understand how Paul and I got from "Hello, how are you?" to "I want to spend my life with you." Bruce asked about our biggest conflicts and how and if we'd solved them.

"We're trying to anticipate the big-ticket items," Bruce explained, "so that when they come up in the future—and they will—they won't blow your marriage out of the water."

Paul looked seasick. Was my Lucida sinking like the *Lusitania?* The Lucidania?

"You seem unsettled," Bruce told Paul. "Unresolved."

"Paul's broken up with me before," I said. "Lots of times. Every time we get too close. He'll say stuff like, 'I love you but I'm not *in* love with you,' or, 'I just can't! Godspeed!' or—"

"So I'm watching *Groundhog Day?*" Bruce said.

"You totally are," I said. "It's so predictable. Then he calls me up, crying, and we pick up where we left off. It's what we do. No biggie, really."

"You can't keep going through that," Bruce said. "It must be tearing you to pieces. Paul? Is this a pattern? 'Marriage, don't tread on me'? Because if that's the case there's no reason to set a wedding date."

"WHAT?" I shrieked.

"I'm . . . kind of . . . criss-crossed and padlocked," Paul said. "I am."

"His sister told him, 'You're too old to make another mistake,' " I said to Bruce. "She's his only sibling and—"

Paul momentarily loosened his grip on himself to reach for the ChapStick in his raincoat pocket. He applied it heavily to his cracked lips.

"A certain amount of caution after a divorce is going to occur," Bruce said. "The degree varies enormously from person to person. But Paul, my friend, you look a lot past cautious. The whole point of my rabbinate is to help people move toward wholeness of being. It's terribly important and a sacred obligation. There's a responsibility that goes along with it to do the very best I can to help you as a couple get off to the right start. I don't just show up and do a ceremony."

"I am locked up," Paul said. "Maybe you could help me unlock myself?"

"Me?" Bruce said. "I'm no rapist."

"What?" Paul and I asked simultaneously.

"I don't go where I'm not welcome," Bruce said. "And it's very clear Paul's not going down that road. You need the kind of help I can't provide. Commitment makes you nervous. Vulnerable. With Gigi, it's just the opposite. I think there's a powerful, mutual love here. I think you both want it to be wonderful. But Paul, you're not ready for a legally binding relationship and all that comes with it. You're not free. Sorry, folks."

I drove the Dino to the Amtrak in New Carrollton to catch his northbound train. We didn't say much in the car. What was there to say?

"I'll call you when I get home," Paul said. He wasn't looking at me.

"Okay," I said.

I pulled away, not waiting for Paul to go inside the station. I didn't want to see him and I didn't want him to see me cry.

"I am no longer affianced," I announced to the 'rents. Mami was already in bed by the time I got home. Big pink cabbage roses engulfed her in a down comforter. Papi was half asleep in the rocker, wearing his striped pajamas, silk bathrobe, and soft worn leather slippers. A movie played in the dark.

"¿Qué?" Mami said absentmindedly.

"What are you watching?" I said, climbing up on the bed and turning on her reading lamp.

"Moonstroke," Mami said. "Ees my favoreet. La familia! Dat Chair cracks me up."

"SHARE," I said. "Her name is pronounced SHARE."

"Right," Mami said. "Chair. Das what I johs said."

"SHARE."

"Den why ees eet spell-ed weeth a C-H?"

"I think Paul and I just broke off," I said.

"Again?" Mami said, not looking up. "Please. Das so re-dohndan'."

"I know," I said. "But we looked at rings in Tiffany yesterday and we went to see Bruce today and then he totally freaked out. We barely talked to each other on the ride to the train."

"Se le va a pasar," Papi said. "El tipo 'sta nervioso, coño. Eso 's normal. Dale un chance." He'll get over it. The guy is nervous, dammit. That's normal. Give him a chance.

"I'm *tryeengh* to watch CHAIR," Mami said, pressing the up volume button on the remote.

As the Brooklyn snow falls down on Ronny (Nicholas Cage) and Loretta (Chair), Ronny tells her, "Loretta, I love you. Not like they told you love is . . . Love don't make things nice. It ruins everything. It breaks your heart. It makes things a mess. We aren't here to make things perfect. The snowflakes are perfect. The stars are perfect. Not us. Not us! We are here to ruin ourselves and to break our hearts and love the wrong people and die. The storybooks are bullshit. Now I want you to come upstairs with me and get in my bed!"

"Dat ees so fabulohs!" Mami said, reaching for her Kools. "I lohvee!"

When the movie ended and Papi began snoring, Mami turned off the TV set.

"Okay, so let me get dees straight," she said, turning to me. "Paul jus' ruin-ed our lives?"

"Yes."

"*Mumita,*" Mami said, gathering me in her arms. She smelled like Bill Blass perfume. "*Mi pobrecita!*"

"I know. I feel like shit."

"Don'," Mami said. "Be pees-ed off eenstead! He was goheengh to be joor husband! One meeneet hees goheengh eento Teefany an' de nex' hees leaveengh Teefany, like, Oopsie, I was johs keedeengh! Wha's weeth heem?"

"I'm trying to sleep, *coño,*" Papi grumbled.

"Paul was goheengh along weeth a dream," Mami continued, ignoring Papi. "A fantasy. An' den, ders notheengh der! Ees like a balloon, one preek, all gone! Joov been through a lot of ohps and downers weeth heem. An' de fact dat joo went to see de rabbi ees very poignan'. Joo went der to talk about marri-age, not to go to sehrveeses!"

"Just kill me now," I said.

"I theenk Paul does lohv joo, doh," Mami said, putting out her Kool and sipping her Diet Coke with Lime through a straw. "Dees has notheengh to do weeth joo or joor behavior. Joo were never bad to heem. Never. He johs got nervous because he knew dees was serious. Ahmbeevahlehn', das heem. He has eeshoos."

"Just kiiill meee nooow," I moaned.

"What he did was not proper," Papi said, eyes still closed. Mami and I both looked over at him. We thought he was asleep! We started laughing. "A guy who is serious doesn't do those things," Papi added. "Paul was getting a bargain. You were more interested and much younger. I told Mami."

"No *tafetán color champán* for me," I said, reaching for one of Mami's Kools. A *Kool.* That's how bad it was. "No Lucida. No little blue box."

"Der ees *tafetán color champán!*" Mami said, handing me her lighter. "Der ees Lucida an' de blue damn boxes! De theengh to do ees dat joo have to get out DER."

"It's not too late for you, *gorda,*" Papi said. "When it comes to matters of the heart it's never too late for anything. You have so much time. No question. Paul's gonna try again, you'll see. It's a father's feeling."

"But I'm a 'hard feet,'" I said, exhaling the wretched mentho-lated smoke.

"I don' theenk joo are such a hard feet," Mami said. "But joo are not for everybody, das for choor."

"A bad, sad day," I said, putting out the disgusting cigarette and taking a sip of the Diet Coke with Lime. Not exactly Parliaments and TaB, but in my angst I'd left my accoutrements downstairs, and at the moment I felt catatonic. "What did you ever do with that wedding list, anyway?"

"I got reed of eet dees morneengh," Mami said. "Absolutely.

Joo want to know why? Because Paul look-ed like a goner weeth hees bagel. I thought to myself, 'Right now, hees a total goner.' Besides, dat lees, de baseeck people weel always be on der, like Manny, so we don' need eet. I can make many lees. Very easily. Ees not a problem."

"Thanks, Mom. Thanks, Papi."

"Papi's sleepeengh," Mami said. "Less leave heem een de rocker for now. Lohv poops heem out."

"Poops me out, too," I said, sliding under the cabbage roses with her.

"*Con* Paul, *no se sabe,*" Mami said, turning off the lamp. With Paul, you never know. "Eet has to do weeth being deevorc-ed an' being cautious an' eh-stohf like dat."

"That's what the rabbi said," I told her, yawning.

"Nighty-nighty, Lulamae," Mami said softly. "An' remember, eef dat Chair can get a man, den probably so can joo!"

CHAPTER FIFTEEN

Flight

*B*asulto warned me, *"No tomes café."* Don't drink coffee. Being Cuban, I drank it anyway. I would come to regret my decision. Stupid when not smart.

José J. Basulto, *Cubano Americano.* Miami developer. Former Bay of Pigs prisoner. Drank his own urine to stay alive in Fidel's jungle prisons. This is who was telling *moi,* a fellow Cuban, *not to drink coffee?* What was *his* problem? My handsome, fearless, no-nonsense pilot, fifty-two years old back in June 1993, was the president and cofounder of the Miami-based search-and-rescue operation known as Hermanos al Rescate, Brothers to the Rescue. The group of volunteer pilots oversaw the rescue of thousands of desperate Cuban *balseros,* rafters, fleeing an island in ruins. The *balseros* navigated the shark-infested Straits of Florida daily in precarious homemade *balsas,* rafts. Many died trying to reach freedom.

The *Washington Post* sent me to Key West to fly with the Hermanos. Mami was a wreckopotamia. About the possibility of *my* death by water.

"*¿Tu 'tas loca?*" she cried. "Dat ees so stoopeed! Dey won' hire joo as a fool-time eh-staffehr an' joor reeskeengh joor damn life for dat fohkeengh newspaper?!?"

"Uh, yeah." Mami had a point, though.

"Den don' tell me. I can't stand eet."

I called Basulto and asked what must I do to prepare.

"Psychologically?" he said.

"Yeah. And any other way."

"You can't prepare yourself for this," he said. "Nobody can. Just be at Opa-Locka [Airport] at 7 A.M. sharp the day after tomorrow as we planned. *Y no tomes café.*"

"Opa-Locka? Isn't that where Amelia Earhart took off on her final, not to mention FINAL, flight?"

"*No tomes café.*"

No one in my family had been back to Cuba since the day Hitler's devil spawn tried to steal my red tricycle, until Papi went on a humanitarian mission in 1990 with several other physicians belonging to B'nai B'rith. Each doctor was allowed to bring along forty-four pounds of giveaway medical supplies. Papi stuffed his carry-on bag with a million samples: Tylenol, Metamucil, Tums, you name it. He brought tampons and Band-Aids and Q-tips. Since he was going for a whole week, he figured that would be more than enough. Every sample was gone in a matter of two or three days. And the poor little Cubans clamored for more like aggressive, extra-needy trick-or-treaters when you've long since run out of Halloween candy, like, "Okay, you've run out of Snickers. Got any raisins, or what?" One weekday Papi was standing out in front of the Hotel Nacional, waiting for his chartered bus, when two very young, very pretty Cubanitas who spoke pretty good Spanglish approached him. They might've been sisters. One looked about twelve, the other was maybe fifteen.

FIFTEEN: *¿Qué tal, señor?* Want company?

PAPI: *¿Qué?*

TWELVE: We like you.

P.: *Gracias.*

F.: Okay, me and her, all day, all night, one hundred American dollars.

P.: What? Oh. Aren't you two supposed to be in school right now?

T.: School. Right. Okay. Me and her, all day, all night, fifty American dollars. What do you think? We do *everything*.

P.: I'm . . . I'm waiting for my bus. Where are your parents?

(The girls have a private summit. Papi wonders what's taking the bus so long. He overhears T. say, "Si, but he hasn't said no yet!" They reapproach.)

F.: Okay, *señor.* Best offer. Me and her, all day, all night, no dollars. Free.

P.: Free? Why free?

T.: You're at the Nacional, right?

P.: So?

F.: You got those little soaps and the toilet paper, right? We'll take those instead.

P.: The little soaps and . . . Instead of cash?

F. AND T.: *Si.*

P.: Why?

T.: We stink.

Papi got on the bus and went to visit the Conservative synagogue where he'd been Bar Mitzvahed so many years before.

It was a dump.

He went to see other places important to him and to our family: the house he grew up in, the temple where he and Mami were married, Baba Dora and Zeide Boris's house, my parents' newly-wed house, which was *our* house, my very own baby guerrilla house.

Dump. Dump. Dump. Dump.

"It was like something out of *Dr. Zhivago*," Papi later said, meaning the scene where Zhivago and his wife return to their old house after a long absence and their once beautiful home is now all gone to hell and inhabited by multiple peasants and other strangers. Ghosts.

"And the food in Cuba was horrible, too, by the way," Papi added. "Forget it, you can only get good Cuban food in Miami. Cuba's not Cuba anymore. Not my Cuba. Not my home. Stupid little island."

Papi was heartbroken. Part of him was sorry he ever went back to witness the ruin Cuba had rotted into. The tropical waste-land he encountered there had replaced the tropical paradise we left behind when we were young, a lifetime ago.

Papi said, "You know what American business would make the most money in Cuba today?"

I go, "Coca-Cola?"

"Nope."

"Mickey D's?"

"Home Depot. The whole fucking place is falling apart."

You can't fight the power. Communists' forte isn't exactly fan-tastic architecture or its maintenance. And yet, in a week or two, Papi turned philosophical.

He said, "Maybe it's good we had to leave Cuba."

This was HUGE. Heresy in the eyes of a typical exile.

I go, "What?!? Why?!?"

Papi said, "Because. Cuba is really a very small island. When we lived there we thought it was the center of the universe. Leaving let us see a bigger world. We learned the world is vast. Much vaster than the one we knew. The experience made us grow."

Unlike so many of our fellow Cubanos in exile, Papi had let go.

To those who insist on romanticizing Cuba, what can I possibly say to change your mind? You're entitled to your romance, even if it is a superficial cartoon fantasy that has nothing to do with me and my people's reality and history. Have your 1957 Chevy Tropicana Bacardi Hemingway Romeo y Julieta cigar cha-cha royal palm tree pipe dream—God knows everybody else does. *Vogue* can go down for "native color" and shoot some beautiful Narciso Rodriguez clothes that cost more than the island's entire treasury holds, maybe put a Ché Guevara T-shirt and beret on a Russian model for fun and stick a cigar in her mouth, like Faye Dunaway in *Bonnie and Clyde.* I won't mind. But I won't go. Not until anyone can go. Until then, I'll cut out the pictures and use them as screen-savers.

This is why, whenever people ask me if I've ever been back and why haven't I, I always think, Back for what? To see what? Cubans suffering under that aging, out-of-it, time-warp gangster who compares himself to Jesus and took my house and Zeide Boris's Camisetas Perro and Zeide Leon's Cuban American Textiles and Papi's Centro Médico Nacional? Or shall I return to my homeland to see all the beautiful little baby whores? Once magnificent buildings in crumbling decay? Silenced dissidents wasting away in jail? Brilliant mulatto musicians on the street who'll never be anywhere else? The talc-white sand at Varadero beach on whose seashell-studded strand Mami and I gamboled? The same sand el Almirante Cristobal Colón and his faithful Jewish interpreter Luis de Torres impressed with their sublime feet? That sand, its strand, those moaning, incandescent seashells, the salty siren-and-shark-filled sea—are all off-limits except to cash-stuffed tourists. Which of these would I care to go and see today? I don't want to be hurt like my father was.

Yet here I was, ready to get into a tiny plane and go toward Cuba. On the day of our flight I awoke at dawn with a catchy Cuban nursery rhyme on my lips: *"Ay, Mamá Inés, ay, Mamá Inés, todos los Negros tomamos café"* . . . "Oh, Mama Ines, oh, Mama Ines, all of us Negroes drink coffee . . ." *Café, café,* give me my *café, coño.* Ignoring Basulto's warning, I polished off a delicious room service espresso—hello, how the hell else was I supposed to be conscious at first light in summer?—and drank a huge glass of ice water. Then I drank a can of *guanabana* (soursop) nectar, ate buttered *tostadas* and a boiled *huevo,* and smoked a Parliament. Yummy. Then I drank one more espresso—this one for the sky, as it were—and another glass of ice water, and peed. (I'm always drinking and peeing, this was nothing new.)

Now I was wide awake and ready to take flight! All right! Though it was already sweltering out at six, I was buzzing with caffeinated alertness and energy on my drive to Opa-Locka. The radio played a lively little Bola de Nieve ditty, *"Espabilate"* ("Wake Up"): *"Oyeme China, tan divina . . . espabilate, espabiiilate!"* "Listen to me, Chinese girl, so divine, awaken, awaaa-ken!"

I climbed into the rear right passenger's seat in Basulto's perfect powder-blue twin-engine Cessna, and my photographer, a guy, sat next to me. Basulto noted the girth of my huge straw tote bag—and it was extra-huge that day—and said, "Hey, if we go down, don't look for your purse, okay?"

"You're joking, right?" I said. Basulto smiled at his hunky young copilot, who smiled back. They both looked like movie stars. Cuban action heroes. "Anyway, screw the purse. I'd dive for my notes! But this *is* a seaplane, right?"

"A *seaplane?*" Basulto said, laughing and playfully punching the copilot's chiseled upper arm. Basulto pulled on a navy-blue baseball cap with the number 2506 in yellow on the front, and aviator shades. *"Niña, if we crash we go down inside the sea!"*

"Inside the sea with the SHARKS?" I said, nervously reaching for my lip balm.

"We have a little storage space in the back," Basulto said, putting on headphones and handing me a pair. "There's a self-inflating raft. If we crash, get in it."

"I'll do that," I said, looking behind me. Yep, there was a raft. Shit, I should've bought a case of TaB and refrigerated it overnight. You could squeeze four six-packs back there easy. You could stack up some cases. I looked out my triangular window as the fierce little plane rumbled and took off, a bird gliding into blue. At 110 mph, we were aloft in a flash, engines droning, sun shining, Basulto singing, "On a beautiful day like today . . ."

We headed south along the 82nd meridian, between the 24th parallel and the Tropic of Cancer. The world beyond the Cessna's clear windshield consisted entirely of bleached-blue sky and white clouds like hunks of torn cotton. Below us the Straits were waveless, flat as a dinner plate. The hot tropical sun sparkled diamonds on the surface that went on forever into the horizon. People always say "only" ninety miles to Cuba. From my aerial view, it may as well have been nine thousand. Or nine million.

It felt sticky in the cockpit. Sweat streaked across and down Basulto's cheeks. Mine, too. My face was wetting my notepad, drop by drop. Thank God I did the waterproof mascara.

"Oye Basulto," I said into my mike, *"porque hay tanto calor?"* Hey Basulto, why is it so hot?

"*Quité 'l aire,*" Basulto's deep voice crackled back to me. "*Come gasolina.*" I cut off the a.c. Eats fuel.

This would be pleasant. Basulto peeled a banana and devoured it, tossing the peel out the window. He had no problem littering but God forbid you mess up his Cessna's seats with your dirty sneakers. So Cuban. Basulto was, as I jotted down in my notebook, stern, affectionate, wry, precise, courtly, spiritual, and fussy. He LOVES planes.

We'd been in the air for almost two hours. I was in massive Parliament and TaB withdrawal mode and had to pee real bad.

"*Oye* Basulto," I said, "*Donde 'sta 'l baño?*" Where's the bathroom?

"*¿i¿El BAÑO?!? ¡Ay, Dios mío!*" He laughed maniacally and shot a macho look at the copilot.

"*Me 'stoy muriendo,*" I wailed. I'm dying.

"*¿Tu tomaste café esta mañana?*" he asked accusingly. Did you drink coffee this morning?

Fuck! Fuck, fuck, fuck.

"Look, I will pay you cash money to land this fucking plane in Cuba," I said. "I don't care. I got Visas, I got MasterCards. We can't be that far away from Havana. I'll handle Castro's troops personally. I've done it before. I have to GO!"

Basulto chomped on a tropical plum, shaking his head and throwing little seeds out the window. So *that's* how you stayed hydrated without drinking coffee or water! This was becoming tragic—for me.

"Hey," my photographer whispered to me, "if you wanna aim at this empty can I promise I'll try not to look at you while you do it."

"Freak," I said batting my hand. "Just . . . go 'way."

"*Oye niña,*" Basulto said, "*mira pa' 'llá.*" Look over there.

I looked out my window. Heat waves shimmered everywhere. We were crossing the southwest reaches of the Straits.

"*¿Mira qué?*" I said. Look at what?

"*Allí. Mira más duro.*" There. Look harder. "*¿Qué tu ves?*" What do you see?

I blinked. Still just shimmer. I found my eye drops and blinked again. It was a mirage, it had to be. It couldn't possibly be . . . could it? It was white and long, like whitewashed Legos vertically stacked in a row in the haze.

La Habana's skyline.

Havana.

Home.

The cockpit was silent. I glanced at Basulto. Was that sweat or tears streaming down his face? Because decades after surviving the Bay of Pigs invasion, Basulto was still enraged. The first man to die in his brigade, during CIA-backed training in Guatemala, wore the serial number 2506. Basulto had painted that number in vivid saffron yellow at the tip of his immaculate plane (and that was the number on his cap). Every time Basulto took to the sky he stowed the memories of that humiliation. When he saved Cuban *balseros*—on that lucky day we found nine live men in two separate rafts—and watched them be plucked from the sea by the Coast Guard (Basulto himself had no legal authority to deliver refugees to American shores), it's like a big FUCK YOU to Fidel.

To me, seeing home for the first time since I left it with my little red tricycle and stuffed lamb was sadder and happier than any words. Mostly sadder. From twelve miles away, the territorial limit from Cuba's coast, I put my finger on the squeaky windowpane and touched La Habana.

With a sigh, Basulto banked the Cessna, turned it north, and flew us back home. To our other home. The one with the bathroom in it.

The Cubans in my life have been in a holding pattern since the Revolution, waiting for Fidel to die so they can alight. I'm really not sure what we're all going to do once Castro croaks, but I think it'll be pretty good and I bet you can watch it on TV. I bet I'll think of how many needles and pins I'll stick in Hitler's demon spawn, that fucker. *Vamos a meter una, dos, tres . . .*

Since the second-most important flight of my life was with Basulto, I'd like to call him up and fly together to a Fidel-free Cuba. And I wouldn't drink coffee that morning, either. Maybe if there's room in the storage compartment next to the inflatable raft, I could stack some TaBs to share with my godmother, Nisia. 'Cause I'm sure they don't have any there. Also, for the eight hundred Jews left in Cuba, who are very poor and haven't been able to get any traditional foods for the Seder since 1960, I would like to bring about one hundred cases of gefilte fish. Excuse me, gehfeelteh feeshy.

Mami would be proud.

Plus I think it goes with TaB.

Acknowledgments

To write a book a Jubana requires a *lot* of TaB and Parliaments. I don't even want to think how many. And a Jubana needs support, because it can get hard. When I appealed to Mami Dearest she said, "I don' know why joor makeengh eet eento such a beeg deal. Every day I watch dat *Today Choh* an' dey have an author on der every damn day. Das five times a week, times a jeeahr, times whatever. De whole worl' has a book."

Sigh.

Nobody knows the trouble I've seen. Except for my kitchen cabinet, that inner circle of people I trust implicitly with my soul and secrets and work. Advisers, friends, psychiatrists, rabbis, journalists, priests, cheerleaders, lawyers, agents, editors. (You might question whether the last three constitute actual "people" in the human sense, and really, who could blame you? But I'm telling you, while my record of choosing fiancés may be, shall we say, pocky, I'm exceptionally talented at choosing everybody else. It's a gift.)

So I have to send a major thank-you note, as it were. Now if I were poor dead Jackie O., I'd have sent it within twenty-four hours. Joo know?

Anyway, here's the director's cut.

These are the five in the backfield, my heroes, my infinitely brilliant and generous quintet:

Jaime M. Naughton, my Irish Catholic guardian angel and former marine. You kick my ass when it and any other part of me resists writing what's too tough. You don't give up on me, especially when I'm at my most give up on-able. We disagree and still like each other afterward. You tell me strongly when the work really works, and gently when it really doesn't. You pay attention. You stay committed. You have integrity. You have grace. You have fun. You have a kayak. You embrace my energy and literary risks. You make me laugh. You make me think. You make me better. You understand me—this is a major miracle. You are the editor and friend of a lifetime. You are my eternal role model. I bow down.

I couldn't live without my beloved *rabino*, Bruce Kahn. My teacher, my touchstone, my friend. You guide my spirit toward wholeness with love, wisdom, humor, warmth, and compassion—not to mention how often you rescue me on Hebrew and Yiddish transliterations, Jewish history, and our beautiful Reform rituals and traditions. I love you and admire you and I will always be grateful you are my rabbi, sent from God just for *moi*.

Manuel "Manny" Roman, you are the world's greatest Bronx-born Puerto Rican psychiatrist-psychoanalyst and friend. Thank you for unraveling my knotty heart with insight and patience and no judgment, and for soothing my seething Jubana mind (what's left of it). You say be brave and strong and Jubanique so I can strap on my ovaries and go for the gusto! You tell me to tell fear to go fuck itself—that's how we solve problems. And you teach me that just because someone smokes Marlboro, it doesn't make him a cowboy. You rule.

Everyone needs a Joe McLellan. Joe, aka Tío Pepe, you were there from before the genesis of *Jubana!*, when I was a wreck-opotamia (even more so than usual). Thank you for getting me down from the tree. Trees. Over and over and over. You're my literary fireman.

And a special gracias to Paul "E.D." Jablow, without whose gifts of a desktop computer and really good peripherals this book would not have been written so well.

In alphabetical order, here are the other most important, trusted, dearest, everything-est people, the ones who also keep me intact and alive, who go far above and well beyond. (And no, you cannot have them for *your* weddeengh guest lees):

- Ana Acle

- Marvin "Gramps" L. Adland

- Lily Anders

- Christopher "Mi Abogado por Siempre" Bolen

- Ken Bookman

- Joan Brciter

- Coca-Cola Company (for TaB)

- Bill Ervolino

- Shanita Furjaníc

- Macarena Hernández

- Robert Kelley

- Joni Mitchell and Leonard Cohen, the bride and groom atop the lyrical cake

- Philip Morris USA (for Parliaments)

ACKNOWLEDGMENTS

- Steve Padilla

- Sidwell Friends School's alumni office

- Beth Silfin

- Anthony Scialli

- Mitch Tuchman

- Sharyn "La Shar" Vane

- Eman "Brastrap Petunia" Varoqua

- Lucinda Williams

Muchas gracias to the inspirational others who also helped:

- Henry Allen (actually, legally, sign-the-mortgagely, Henry Southworth Allen IV)

- Suli Baicowitz

- José Basulto

- Katherine Beitner

- Anita Bernhaut

- Helen Buttel

- Michelle Dominguez

- Kimberly French

- Joel Garreau

- David Gonzalez

- Jonathan "Johnny" Gordon

- Peggy Hackman

ACKNOWLEDGMENTS

- Mary Hadar

- Michael Hill

- Larry "Mumo" Katz

- Jean Marie Kelly

- Bridgette Lacy

- Lewis Lawson

- Maureen Lewis

- Zev Levin

- Elizabeth Llorente

- Phil McGraw

- Tom Miller

- Emanuel Ory

- Michael Pakenham

- Rem Rieder

- Laurie Rippon

- Richard Rodriguez

- Joe "Papo" Vidueira

- Neville Waters

- Richard Wertime

- Daisy Wise

Mil gracias to:

- My exceptional agents and fearless leaders Jane Dystel and Miriam Goderich, the Jewish and Cuban literary dream team. You saw something in me before I could see it in myself, and smoothly drew me out from under de bed and into a real live book dream. Wow.

- My family, particularly my beautiful *mamá*, Ana Anders, and my *tío*, Bernardo Benes, whose memories, feelings, and observations helped immeasurably to feel een dc hohls of de heestohrees.

- My reclusive Republican vegetarian copy edtor, Eleanor Mikucki. Eleanor, who knew a reclusive Republican vegetarian could be this good?

- My senior production editor, Sue Llewellyn. Sue, I still have no idea what the hell you do, exactly. But maybe that's because you make it look so effortless.

- My Rayo editor, René Alegría. What can I say, René? You had me at "Hello Kitty."

- My other Rayo editor, Andrea Montejo. Someday, dear, we're getting you some how-to-be-mean lessons. Just not during my lifetime.

I love you and thank you all.

bonusPAGES

© Charles Gupton

GIGI ANDERS, a *Washington Post* special correspondent, was born in Havana, Cuba. She has written for, among others, *Glamour, Allure, Latina,* and *American Journalism Review*.

A CONVERSATION WITH GIGI

The word "Jubana" is interesting. Sounds like the name of a super-hero.

Superheroine.

What's your Jubana superpower?

Well, being a Jubana, there is always more than one super-power. I contain multitudes. To paraphrase Peggy Lee, I'm a *Jubaaaaaaaaana*, J-u-b-a-n-a. I'll say it again: I got an American Express card says there ain't nothin' I can't do, I can make a killer outfit out of Isaac Mizrahi for Tar-jay and I can make a Jubaaaaaaaaana out of you. 'Cause I'm a Jubaaaaana, J-u-b-a-n-a. And that's all.

So just what is a Jubana?

A Cuban-born Jewess.

You've always lived your life on the fringe and your adolescent friends seemed to reflect this.

My current ones do, too.

Were you the quintessential fag hag in high school?

Please. There were two gay guys—that we know of—in my tiny class. And yes, I was close to both. But I'm attracted to all kinds of interesting people. I'm a Catholic hag and a lesbian hag and a black hag, too. Sometimes all in one person.

What do you have that Margaret Cho doesn't?

Jubano parents. Spanish as a first language. The ability to make a devastating black bean hummus.

You seem to have a propensity to simultaneously feed into yet refute certain cultural stereotypes about Jews, about Hispanics, about even being a strong, intelligent woman.

Gracias.

Is this type of internal push and pull with yourself exhausting, or do you find it the best indication of what drives you?

Both. It *is* exhausting to be *moi*—I burn up of a LOT of energy every day. That's why I require 16 hours per night to recover from myself. On the other hand, I do like a fast, intense, abundant, colorful, joyous ride. Too much caution, like inertia, sucks all the fun out of life.

You have a very emotional passage in the book about your first sexual experience. How do you explore something like that, as an adult?

Very carefully. It was hard. The hardest writing I've ever done.

What was it like to write about something so personal, so human?

Scary. Mean. Embarrassing. Painful. Liberating.

So tell me about your ultra-special Mami Dearest. She's quite the radiant glamazon.

She is. That complicated, gorgeous gal takes a lickin' and keeps on kickin'! She refuses not to be fabulous at all times. We've had our clashes, God knows, but I'm glad she's my mother and I love the twisted, crazy bitch. I mean that in the best sense.

Like her, can you, too, put on makeup, smoke cigarettes, and watch Old Movies at the same time?

Oh, that's what I do every day! I've been trained from birth by the master Mistress.

Is this what you call Jubana multitasking?

To a non-Jubana, I guess it is. For us, it's normal. Women in my culture—okay, me and my mother—are like feminine octopi. We got a lot of things going on at once all the time. Like, if we're smoking, we simultaneously have the Tic Tacs or the espresso or the TaBs happening. If we're doing the toenails, we're also watching *Now, Voyager* and talking to our girlfriends on the cell. If we're doing a home bikini wax—well, okay, I take that one back. You can't do more than bikini wax if you're bikini waxing. But other than that, we're stimulating because we're stimulated, and vice versa. Maybe too much so. It's from years of Cuban hypercaffeination and too many sessions with too many Jewish shrinks.

SNAPSHOTS FROM A LIFE

Mami and Papi during their 1950s Cuban court-ship. I know. They're Hollywood-gorgeous.

As a baby, my all-time fave look was opaque black tights (as in the birthday pic)—*y nada mas*. And nothing else. Except the jewelry, *por supuesto*. That's my Mami holding me in the *sala* of our lovely Cuban home.

Loved my scrapbook.

Love birthdays. Especially Cuban ones. You get to shake your maracas and wear bespoke black velvet dresses with icy-pink satin trim. So chic! Unfortunately, being Jewish, my mother ruined the dress's exquisite lines by forcing me to wear a cashmere cardigan on top. (My birthday's in December, and in Havana, 75° is considered arctic.) That's my little cousin Joel next to me. This was taken on my second birthday.

Loved my milk. Loved my tights. Loved my mother.

My glamorous parents, David and Ana, and me in our backyard on my second birthday. Mami is pregnant with my sister Cecilia, her second child. This is a bittersweet photo; Cecilia died only a few days after her birth.

I've always loved riding the painted ponies. Painted ponies, painted ladies. Cubanas would never DREAM of leaving the house without full makeup. Doesn't my mother look like a redheaded Grace Kelly? And note her perfect 19-inch-long fingernails. That is a profoundly Cuban look, circa 1959, which has, unfortunately, never ebbed.

When Bangs Grow Out.
I was always being asked to be the
flower girl at Cubanos' weddings.
I was good at petal sprinkling. I was
two years old in this Havana pic.

I look happy because I'm on a painted
pony. But this is actually one of the
saddest photographs my family owns.
It was taken at Havana's José Martí
airport on our last day in our beloved
homeland. There would be other
ponies. But not like these.

After the pony ride, I hugged Mami's
best friend and my beloved godmother,
Nisia Agüero, to say *adios*. Nisia
stayed in Cuba by choice. I've never
seen her since.

El exílio. Mami and I are relaxing in the yard of Las Casitas Verdes, the Miami Beach apartment building we lived in for six months right after we fled Cuba. We're smiling because (a.) *I love* wearing Mami's eternally chic, big black Jackie Kennedy sunglasses, and (b.) we still thought Jackie's handsome husband, *el presidente*, would solve our problems via The Bay of Pigs Invasion and we could go back home. We were wrong. But not about (a.).